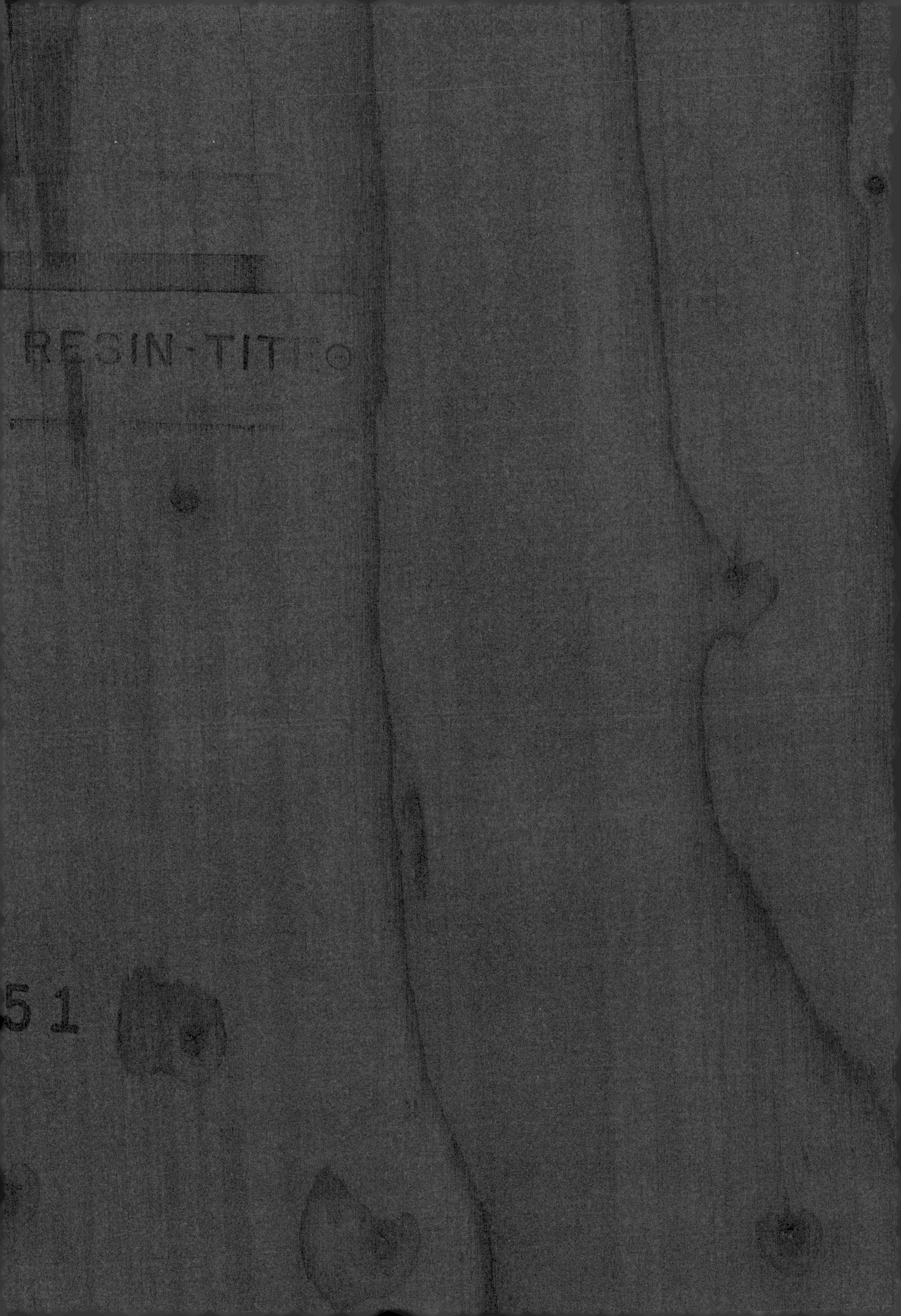
RESIN-TITE
51

11/98

TOOLS OF THE TRADE

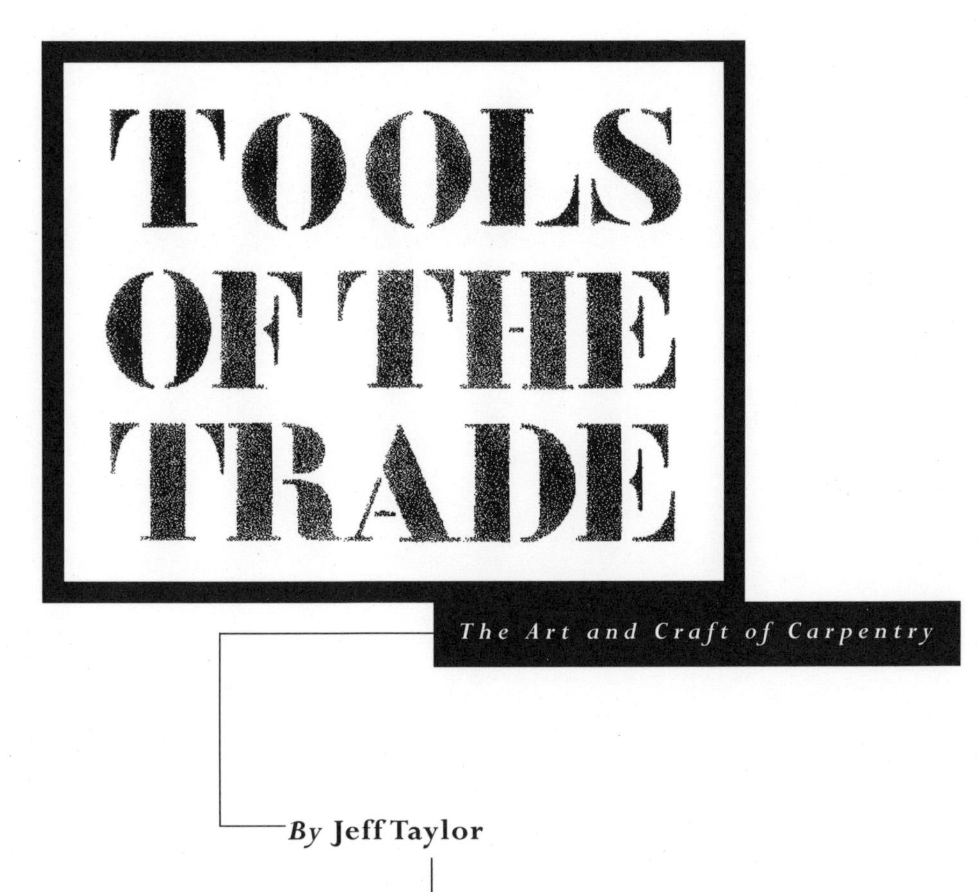

By **Jeff Taylor**

Photographs by **Rich Iwasaki**

CHRONICLE BOOKS
SAN FRANCISCO

FOR JOY and SERENITY

The essays in this book have appeared in slightly different form in *Harrowsmith Country Life* magazine, except "Sawhorse: Nonpareil Trestles," which appeared under a different title and in slightly different form in *Country Journal* magazine.

Design by Jane Elizabeth Brown.

Cover photographs by Kevin Ng.

Library of Congress Cataloging-in-Publication Data:

Taylor, Jeff (Jeff Damon)
Tools of the trade: the art and craft of carpentry / by Jeff Taylor.
p. cm.
ISBN 0-8118-1273-1
1. Carpentry—Tools—Anecdotes. 2. Carpentry—History. 3. House construction—Popular works. I. Title.
TH5618.T38 1996 95-47428
694'.092—dc20 CIP

Printed in Hong Kong

Distributed in Canada by Raincoast Books
8680 Cambie Street
Vancouver, B.C. V6P 6M9

10 9 8 7 6 5 4 3 2 1

Chronicle Books
275 Fifth Street
San Francisco, CA 94103

Acknowledgments

ALTHOUGH NAMES HAVE BEEN CHANGED THROUGHOUT THIS BOOK, THE AUTHOR IS GRATEFUL TO THOSE PROFESSIONALS WHO GENEROUSLY SHARED THEIR KNOWLEDGE OF WOODWORKING AND HOMEBUILDING. THEY WILL RECOGNIZE THEMSELVES, NO DOUBT.

OVER THE YEARS, THERE HAVE BEEN MANY OTHERS WHO HAVE GIVEN HELP, FRIENDSHIP, AND ENCOURAGEMENT AS THE AUTHOR HAS LEARNED TO BUILD AND WRITE, AND HE WISHES TO THANK THEM AS WELL: JIM MASIMER, TOM CARPENTER, LOWELL BELL, MARK AND SANDI ERICKSON, BONNIE AND BRENDAN CARMODY, DICK JALBERT, CHARLES MILLER, JIM GUEMMER, SARA PACHER, JERRY MINER, DAVID M. CAMP, TERRY KRAUTWURST, BRUCE WOODS, BOB MUTZ, JON ADOLPH, JEFF WILLIAMS, JACK HART, DAMON KNIGHT, KATE WILHELM, CAROL DEPPE, FLORENCE GRAVES, PAT STONE, ED RETTIG, AND JERRY BOAL. SPECIAL THANKS TO HIS AGENT, JEAN MACDONALD, AND TO JON TAYLOR, WHO CHECKED THE ORIGINAL MANUSCRIPTS, AND TO JAY SCHAEFER AND KATE CHYNOWETH AT CHRONICLE BOOKS FOR THEIR SUPPORT, ADVICE, AND GUIDANCE, AND TO BROTHERS TOM AND TIM TAYLOR, WHO HELPED BUILD THAT DAMN HOUSE.

CONTENTS

1 *Introduction*

7 **HAMMER** ✦ *How to Bang Wood*

17 SAWHORSE ✦ *Nonpareil Trestles*

25 **TOOLCHEST** ✦ *The Grown-up's Toybox*

31 STRINGLINE ✦ *Getting Straight*

37 **TAPE RULE** ✦ *Measure for Measure*

43 LADDER ✦ *Getting High*

49 **TRANSIT** ✦ *Shooting the Invisible*

55 KNIFE ✦ *The Penny Knife*

61 **HAND SQUARE** ✦ *A Matter of One Degree*

67 PLANE ✦ *The Plane Truth*

73 **HANDYMAN** ✦ *The Bequest is a Question*

79 FRAMING SQUARE ✦ *Carpenter's Calculator*

85 **HAND DRILL** ✦ *The Original Cordless*

89 **DRYWALL TROWEL** ✦ *Master of Plaster*

95 **HANDSAW** ✦ *Why, Daddy?*

101 **SLICK CHISEL** ✦ *An Auldie but Guidie*

107 **PIPE WRENCH** ✦ *Of Steel and Lead*

113 **BLOWTORCH** ✦ *Quest for Fire*

119 **CROWBAR** ✦ *The New Barbarians*

125 **DRAWKNIFE** ✦ *Estate of Grace*

131 **CHISEL** ✦ *Keen Edge, Steady Hand*

137 **BENCH VISE** ✦ *My Only Vise*

143 **AWL** ✦ *George on My Mind*

149 **KEEL AND PENCIL** ✦ *Making Your Mark*

155 **SPIRIT LEVEL** ✦ *Centering*

163 **YANKEE SCREWDRIVER** ✦ *Farewell to the Old*

INTRODUCTION

Ideally, a person should build one good house and live in it: a building that will probably outlive its builder, a structure that takes on a life before anyone moves in, a home where bills are paid on the kitchen table and babies made on the hearth rug, a place of gentle memory to sons and daughters. Do it once, and you may want to do it again, and again. The power of building houses, using only the tools in your hand, can take over your life.

For instance, I wanted to teach English on some foggy old campus, sitting in an ivy-framed window and taking young minds aloft on the winds of language and the wings of literature, smoking a briar pipe all day, mumbling and slapping my tweeded knee at Chaucer's filthy jokes.

But I became a carpenter instead, learning the lore from a dying breed, the last of the tinkerers and alchemists who once populated the field of carpentry. There was Morris, blind in one eye from a collision with a police vehicle that killed him, resuscitated for ten more good years of carpentry; and Dick, who so fixated on rooftops as a young man parachuting into Normandy on D day that later he could instantly draw and calculate the roof system of the most complicated house plan ("I just close my eyes and pretend that Germans are shooting at me."); Phil, who hated the United States government for its crimes against the working class, and who aspired to be the chief scaffold engineer when Washington fell; and first and wisest of all, Swanee, my sainted boss, a quiet man who undertook to teach me the rudiments of what he knew. From these and other geniuses, I learned about carpentry, woodworking and construction tools.

In the beginning, all of my tools fit in a shoe box. To be an adequate carpenter, I had to buy a hundred basic tools. To work wood and build cabinets, I reluctantly purchased even more. Fixing mechanical systems, such as plumbing and electrical wiring, required special tools, and I acquired them. But at some point, I began to collect them for pleasure.

My journey began in 1970. There were still carpenters alive who had built entire houses using only hand tools. Swanee, the son of one, by then an old man himself, took me on a few years before he died. His tools kindled strange passions. Just by pulling a plane from his toolchest, my first employer opened up entire worlds for me to explore. Words had long been tools to me, but now I touched rosewood and hornbeam, brass and bevel gears, tangible, lovely *things* in my hand. It was inevitable that I would write about them someday. Now when I use hand tools, it comes back: the settings, seasons, problems, places, and people.

According to Swanee, the craft was about to enter its most debased time, a dark era of assembly-line construction when much of the hard-won knowledge about woodworking would vanish. He spoke of days to come when a giant machine might squat over an excavation and excrete an entire plastic house. The only job for a human then, he warned, might be to hand this lumbering carpentroid a wad of building paper.

A single day of actual construction is worth any number of books on how to do it. No manual can tell your hand what a hammer can teach it; sentences that only call out the placement of 2 x 12 joists don't describe how the weight grinds into your shoulder when humping them from stack to structure. The simple words "blood blister" are painless to read. By contrast, an actual smashed thumb or two will teach you volumes about hammers.

I've spent years thinking about this book and working on its chapters. In the interim, I've learned a little about writing. I've also noticed how fast the technology of carpentry is changing. Tools and materials keep getting better, more expensive, and to some extent more dependent on technology. If my fairy godmother ever gave back my twenty-first birthday, I'd have to learn a lot about nail guns to survive on the modern construction site. When I started, early versions were just beginning to proliferate. Swanee could have been right: by the time the nursing home attendants see me setting invisible nails and pounding them without a hammer, robots with nail guns may have replaced carpenters entirely.

But don't bet on it. Humans like to build. They like tools. Everyone has a dream house in mind.

This book is about hand tools, ladders, sawhorses, safety, and methods of construction, and about the people who taught me about them. It is based on over twenty years as a professional woodworker. It takes that long to understand how little any one person can know about the immense body of knowledge, much of it already lost, that is carpentry. At one time, an architect *was* the master carpenter, as the Greek etymology implies. Now they are two jobs, with entirely different work environments. The architect is the one wearing the tie and pristine hard hat. The carpenter is the one with the tools, the person who actually does the work.

I have a certain automatic screwdriver, a small but precious Yankee, given to me one winter night by a retired cabinetmaker named Joe. We had been discussing the albatross of sentiment that forces us to keep outmoded tools, old gadgets that would still do the work if only there were time enough. He had boxes of them. "Nobody cares anymore about good work," said Joe, flipping another log into his shop stove. Flames licked out as he slammed the iron door. "We got so much leisure time that we gotta hurry to get the work done and get our asses home. You watch. Carpentry's gonna go straight to hell."

I am happy to report that Joe was both right and wrong about the future of carpentry. Good work is rare, as always, but still in demand. In addition, there are tools and devices available now that are better and more accurate than any manufactured during the golden age of carpentry (1865–1955). Joe forgot about art. The recent trend toward sybaritic bathrooms and breezy country kitchens derives, I believe, from purely artistic impulses.

Among other things, art is the pursuit of good health. If the Victorians understood architectural ornament and the beauty of good tools, they sorely needed the concept of central heating systems connected somehow to all bath areas, and kitchens that didn't resemble closets. Anytime I'm tempted to romanticize the olden days of carpentry with rosewood fretsaws and toolchests full of molding planes carried on buckboard wagons, two words crush that reverie: water heater. At the turn of the century, you could die from taking a bath, and only robber barons had saunas. Children ate oatmeal in dark corners and squatted outdoors, the servants had TB, and everyone took laudanum to stay sane. Sure, the Painted Ladies were lovely houses built by wood wizards, but structurally and mechanically, new houses today are built better.

They are not necessarily built by people who love the craft. About 1987, I noticed that all my employers were trying to kill me. It was not entirely their fault; I was past thirty, old for a carpenter, and their bottom line was always speed and profit. No one encouraged the use of hearing protectors or eye protectors, and hard hats were definitely optional on residential building sites; in fact, any carpenter wearing a hard hat would have received a second look for trying to impersonate an architect.

These attitudes were not conducive to my longevity. I found my last job on a commercial construction site. Everyone wore hard hats, but no one was encouraged to use hearing or eye protectors when using saws or hammers because it ate up precious fractions of a second to put them on every time; using valuable time protecting your own expendable hide was like stealing money from the general contractors.

One day an OSHA inspector came on the site and burned through an entire book of tickets: the compressors had no belt guards, the power saws had bare wiring at the plugs, the heavy drills weren't double-insulated, the ladders were improperly rated, there were no sawhorses, the power cords had never been inspected, scaffolding was absent, several workers had already been injured—all in all, coming to work here was like testing artillery shells with a hammer, said the OSHA inspector. You might not get hurt right away, but it wasn't safe according to strict government standards set by the Babylonians and updated since.

"Yeah, but it's turning a profit," said the owners, who paid the fines out of petty cash and overrun money.

I escaped. Not before my neck was permanently torqued and my hearing impaired like a rock drummer's, but I got out of the carpentry-for-hire field and into the lucrative and fun writing industry. I only starved for about eight years, and here you are reading my book about tools of the trade.

It takes a while to find the *meaning* of tools, the aura of them, if you will, the way they seem to be asleep until you learn how they work and how to use them. Suddenly you have a tool that is yours, and more than yours, because it has a history that precedes your ownership. It may have been handled by giants and wizards of the craft; it may act a little skittish in your hands, but at that moment, you are becoming part of its working life.

In learning about tools, I may have discovered the core of carpentry. It's something that doesn't change over the course of millenia; it's the same intelligence that informs the building of earthen huts and the Taj Mahal. Many call it craftsmanship, but that's not it either, and not only because it's a sexist word. Better not to name it at all. It will be the same ineffable whatever long after the last tree is cut down and we build homes from spun petroleum. Surely the soul of carpentry will be there in the building of space stations, if humankind gets that far. Should those structures leak vacuum, we'll know it was missing.

If you can enjoy the process of building while simultaneously attending to minutiae and the big picture, if you know what you're doing from long experience and careful thought, if you can fill the unforgiving minute with sixty seconds worth of rise and run, if your tools and distinctions are sharp, then the result will please you and anyone who knows what to look for in a well-built house. But you still won't know what to call it. Pride of workpersonship? The name is not important.

You can build a hundred more houses, knowing that you'll continue to raise them in your mind after your body fails. At a certain point, upon a day, you almost become the work, a moving and cognitive part of the tool in your own hand. Only for a few seconds, a mystical space that goes away too soon.

But these are the silent moments of revelation. Listen hard, and you will know why you wanted to learn how to build, and what hand tools can teach you.

THE FRAMER
HICKORY
U S A
HART TOOL CO

HAMMER

How to Bang Wood

Technology giveth. If the course of your life requires you to drive a large number of nails, as in the building of a new home, you have a choice: use air-driven nail guns or the traditional hammer. The nail gun is quicker and therefore cheaper once we accept that time is money, money is power, power is good, and more is better.

But even under the batlike shadow of my forty-fifth birthday, with precious time slipping through my fingers, I favor the simple empowerment of a hammer. It is not solely the aesthetics of modern compressors and nail guns, or the deleterious effect of these on young crews who playfully depress the safety and shoot nails at each other—"hunting Pinocchio," as it's called. If you like speed, they can bang you up a house to the gentle strains of Twisted Sister and staccato *ka-chots,* and they'll do it much faster than grotty old quartogenarians with their many hammers and tuneless sobriety. For the developer, pneumatic power nailers mean numerically maximized dwelling units and minimized workforce redundancy: fewer carpenters, for a better tomorrow. Technology taketh away.

But on the downside, I've seen awful things and heard true horror stories: an overlooked front door, frame and all, falling out of its unshot opening when a potential buyer closed it, or cabinets that came crashing down, to the detriment of crystal and china, because someone forgot to put the correct nails in a gun. In the bay-front town where I used to live, an entire condominium roof, shingled with air-driven staplers, was deshingled by a windstorm a week later, and in the next gale stripped once more. The third try, put on with hammer and nails, took twice as long to shingle. Profits suffered. Arms ached. But the roof held, and abideth.

And by the way — unplug the compressor and see which works better without a nuclear power plant behind it. I know how to repair hammers, of course, and can even field-strip several models of nail guns; I am, however, a little vague on whether the control rods go up or down in a breeder reactor. (Complexity requires a perfected expertise, as they used to say in Chernobyl.)

Which is not to say staples or nail guns are inferior per se. The condo roof was human error, a little too much air pressure that drove the staple most of the way through the shingle, and too few hot days to seal down the tabs. Used correctly, the general effect is almost the same; in fact, the air-driven fastener is perhaps a little stronger, certainly far faster. No cabinet shop or production-carpentry crew should be without them.

But the hammer, in all its forms, is more versatile and rewarding because it changes the way you think. Purpose dictates design, as in the mechanic's ball peen that will install a rivet, drive a cold chisel, and work sheet metal, or the way an upholsterer's hammer holds a tack magnetically. In carpentry, a hammer—the same tool that raised the topless towers of Ilium and rebuilt Hiroshima, cousin to the fist axe that was humankind's first tool—will require something more of you than a trigger finger. You will pay in the coin of sweat, sore muscles, and calluses. With practice, you find the right coordination: place the nail, set it, move fingers to safety, and drive it home, missing the rhythm occasionally and then, like Thoreau, keeping your accounts on your thumbnail. You earn, eventually, a skill.

After twenty years, I am still learning all the ways to use a hammer, sometimes haunted by the thought that I have been overlooking yet another way to use a hammer—probably something obvious—for two decades.

Upon a day, I became a carpenter's helper. It began as a summer job in 1970, another paycheck between semesters to finance, ironically, my education. Burdell Swanson, a second-generation carpenter with thirty-five years in the craft, was my first teacher; at nineteen, I was his last apprentice. After a quick interview in which I was exposed as a fool, we began the summer by renovating a Victorian three-story, a modified Carpenter Gothic with bartizan turrets and fancy scrollwork from plancier to pedestal. He showed me what he wanted done, and I did it.

At first, Swanee simplified his lessons, presenting tools in their conceptual form: a saw reduced the wood to parts of a certain length, and a hammer spliced them back together in a pleasing order. My boss had an assortment of good hammers in a variety of weights, lengths, and styles, most with hickory handles and, strangely, all with straight claws instead of the deep curve that pulls out nails so well.

Fortunately, I already knew how to use a hammer: one puts the pointy end of a nail down, and with that tool, one whacks hell out of the other end. This was encyclopedic knowledge compared to my comprehension of the square, the dryline, the transit, the handsaw, the nailset, the chisel, and all other tools except the pencil. But with the hammer, I felt I was on safe ground.

However, my employer demanded proper nomenclature, correcting nearly every word out of my mouth. It was, indeed, called a hammer, but the right name for the hole in the head was "adze eye"; the side of the head was the "cheek"; the knobby part was called the "poll," not "the knobby part"; and the actual striking surface was called the "face" in preference to "whanger." The face could be flat ("plain") or slightly convex ("bell-faced") to drive a nail slightly below flush. "Hit the sweet spot," he'd say, touching the face, "just like a tennis racket. You took tennis in college, right?" On rough work, where hammer marks wouldn't matter, ricochets could be avoided with a cross-checkered (milled) face, a feature found on all rough framing, drywall, and roofing hammers. He frowned on the term "waffle iron," but he had heard it before—"Yeah, *framers* call it that." It was the job description he abominated, not the individuals.

The first month, I assumed he had hired me merely as an outlet for verbal cruelty. He never raised his voice or lost his temper; he might simply observe that a blind person could tell how bad my work was by the way his seeing-eye dog recoiled. In a few terse words, he would conjure up all the trillions of nails driven in human history, some of which had been driven perfectly. Unlike those hammered by summer dilettantes, he implied.

I had purchased an inexpensive chrome steel curve-claw with a rubber grip, feeling that proficiency derived from the use of a single hammer for every task; it seemed obvious to the meanest intellect. Furthermore, after two weeks of constant practice, I could twirl my little 16-ounce nail-beater on the claw and return it to its metal loop on the downswing, in the manner of the Lone Ranger. Swanee had never mastered this manuever, apparently.

One day while we were installing new shiplap siding, I noticed that he was using a different hammer, a little heavier and longer than before. Without thinking, I did my flying roll, and he shook his head, his face as unreadable as Tonto's. "All right, Wild Bill, why don't you drive this nail in for me," he said, pointing high on the siding. He handed me an 8-penny siding nail. "But do it with one hand."

I had some difficulty. He watched, scowling. Finally: "Here, gimme that," he said, snatching hammer and nail. He put the nail head against the cheek of the hammer, holding it between two knuckle joints, the hammer head in his fist. Then he slammed it against the board, where it stuck; he flipped up the hammer and drove it with three expert whams, leaving the nail head one-sixteenth of an inch below the surface. "Like that," he said. "A useful trick."

(Learning to nail one-handed, I found out, is not just for showing off. Much later, I saw carpenters performing all sorts of acrobatics, such as propping a heavy fascia on the scalp and blindly setting the nail overhead, that this method obviated. With a hammer in one hand and a nail in the other, the average human being has no hands left to hold the work.)

The lesson wasn't over. "Now show me how to find the next stud over from that nail, without measuring." The studs were sixteen inches on center, hidden under sheathing. I'd been using my tape to locate them. I dithered.

He snorted and slapped his hammer flat on the wall, head touching the nail. Without looking, he placed another one at the end of the handle. "My hammer is sixteen inches long," Swanee said, wiggling it in my face. "Yours is thirteen and a half. Another useful thing to know about a hammer."

I never twirled again. In return, he began to pass along the finer points. He was capable of carpentry at invisible levels, where the nail head vanishes under a chiseled-up sliver; not just countersunk but sunk without a trace, like a U-boat. The trick was to leave no hammer-sign at all. Even hammer-dented wood, as Swanee demonstrated, could be undented by applying a wet cloth and steaming it with the tip of an electric iron.

Soon the rubber grips on my hammer had done their work, imparting to my paw the most magnificent stinging blisters. This opened the subject of wooden handles.

Swanee showed me how he had bored out the bottom of his wooden handle almost to the head and filled it with castor oil, which bled out into the wood to make a microscopic slick on the surface; hence, fewer blisters. (Years later, I found out that Edgar Cayce had suggested warm castor oil as a rub for arthritic joints.) By capping the opening with beeswax, my mentor could dip screw threads to glide them into pilot holes.

In theory, his oil-soaked wooden handles would last forever, largely because he didn't pull nails with them. "By me, the claw's a dumb way to pull finish nails," he said. "You gotta use a block to keep from scumming up the wood. I like a straight claw. See, it balances better. More weight directly over the face when you drive a nail." He used plierlike end-cutters, which he called "nippers," to remove small nails. "One tool, one hand, more control. You can girdle a big nail and snap it off flush, and cut the little ones."

For framing, he used a small crowbar and its tiny cousin, the nail-puller or "cat's-paw," for big nail extraction. His framing hammer was a mere 24 ounces, and he had an identical one with a sturdy fiberglass shaft for demolition work. (For that task, he wore gloves anyway; still no blisters.) The straight claw was ideal for prying apart boards, and with a fiberglass shaft, he could yank out nails and clean old boards like a machine.

Warming to the subject, he expounded on handles: which made the best blisters (rubber), which handles transmitted the most shock (steel), which gave the least (fiberglass and wood), which heads had a hardened center and soft steel around the rim of the face to avoid chipping (the good ones, like his), and which ones could shatter and flick out an eye (cheapos, like mine).

That weekend, I bought three more hammers, exactly like his, a bottle of castor oil, and a nail-puller. My lifelong study of hammers had begun.

Thereafter, I watched him with new eyes. When an expert nails a board, the hammer fine-tunes the placement: maybe a few light taps outboard at the top to close a miter joint, for example; and then the nail is set and driven. He was a wizard with the nailset, knuckling it between the last two fingers of his left hand as thumb and index held the nail. Deployment involved a simple fingerwalk into position, an eye-blink later. It made more sense, he said, than setting the nails and going back to countersink them.

Nails were an entire subtopic of hammering lore. The diamond shape of a nail's point can act as a tiny wedge, causing splits in thin wood; he showed me how to blunt the nail by rapping it with the cheek of the hammer—"Don't mar up the face with nail points"—to crush the wood fibers as it penetrated. Or nippers could also amputate the point. For hardwood, in lieu of a pilot hole, carpenters often lubricate the nail shaft with beeswax, hair grease, even spit.

When we got to the cabinets, he produced a cross peen, 12 ounces of truly weird-looking hammer, and suggested that I needed yet another tool. The peen would drive tiny nails into the tightest spots, and he proved it. With misgivings, I laid out the money. (Good hammer; still have it.)

On the roof, the seminars continued. "Now this beauty . . ." was the AJC hammer, devised by A. J. Crookston as a modification of the old roofer's hatchet, designed for asphalt 3-tab shingles. The peen incorporated an adjustable gauge and a tiny shingle cutter with a replaceable blade. Did I not agree it was exquisite? A fifth hammer, another dent in my hammer budget. For cedar shingles and shakes, he preferred a light roofing hatchet; but he owned a rubber-handled slater's hammer as well. Rubber? But I thought—

"Doesn't slide off slate." Oh.

Roofing tools lived in his pouch, but nails went in a little box on his chest, a "stripper"; with two fingers, he could slide out four nails at a time, point down. It had two slots in the bottom and four tiny spring-loaded gates, and is still commercially available.

So far, we had done minor framing, interior finish, cabinetry, and now roofing, each requiring a different hammer. Swanee deplored the breakdown of construction into a multiplicity of specialties. Although it made the work more efficient, it also made it more mechanical; the trend toward mass production degraded the sum of knowledge and sapped the most vital component of craft: the human spirit. "Learn it all, everything you can," he told me. "Otherwise, you're just another robot, banging nails."

That onomatopoetic sound was his favorite word for what he saw as the nadir of the craft: tract-house production carpentry by framers. We cruised past a development once, and the sight sent him into paroxysms of disgust. "Those damn houses are just slapped up and banged together," he muttered. I thought the concept of an air-driven nail was rather nice, but kept silent. "Promise me you won't turn into a framer," he said. "Not while I'm alive to see it."

We worked together a few more years, but the day came. I mourned, but Swanee had said to learn it all, and in that spirit I joined a framing crew. That fine old man had imparted a special meaning to the word "carpenter," and I tried to enhance my education without being corrupted. There was an odd pride in being able to build stair carriages or to lay out rafters, complicated mental work, using knowledge handed down by Burdell Swanson, master carpenter, while working among people he would have called bad company.

It was an incomparable time, swinging my hammer all day long and learning how to make the skeleton of a house; on some of the hottest days, with ice-rimmed beer waiting at the end, I even contemplated a career as a framer, until I realized that I had seen no old framers. The work was fun, but brutal. When it rained, we would don oilskins and keep working, making turkey-gobble sounds after another animal too dumb to go inside.

In the Rockies, I worked for Hal, the Knight of the Blazing Oath, whose common language transcended the merely obscene. Swanee seldom swore, other than occasional analgesic dammitry, but Hal never stopped. Upon encountering a mistake or obstruction, he favored it with filthy metaphors or his workhorse, the All-Purpose Noun and its variants.

He shunned air-driven nailers only because he hated the hoses. Early in his career, he had tripped on one and fallen off a roof, doubtless cursing them all the way down. He valued his health more than gold; except when hammering, he wore ear and eye protection because his father had gone deaf and his uncle had lost an eye in the trade. "I should ** wear goggles for *** nailing, too, but I ** don't. ** it, right?" Wrong. Price an eye sometime. Still, who wears safety glasses for driving nails, except the one-eyed?

When Hal wasn't framing, his secret passion was building log cabins in the mountains, and I worked with him there on weekends. His hammer for both purposes was a framing hatchet, customized to his dual needs. He used a flat-sided, offset hewing hatchet with a milled face; and although he was a mighty framer and exacting log builder, all nailing problems succumbing to sheer force, his visible woodwork —verge boards and fascia, which form the perimeter of the roof—suffered from what his employees called "the union label," a circular or half-moon mark with a distinctive waffle pattern. Swanee would have used a bell-faced 20-ounce on fascia, so I did.

Framers, I found, had developed a few hammer skills of surpassing practicality. When we were hauling heavy 4 x 8 sheets of underlayment up flights of stairs, I noticed one man using the claw of his hammer to hook under the bottom edge, providing a convenient handle. I tried it, and instantly adopted a new trick. I also discovered that four 8-penny nails held a toenail better than sinkers, which tended to split the studs unless dulled. Someone showed me how to bend a 16-penny sinker slightly on the claw of my hammer, so that when toenailing parallel planes, the nail curved down the center of the board rather than piercing through the other side.

Each day brought more knowledge. I learned that a 32-ounce hammer was, in effect, a baby maul that would cinch up the heaviest framing members, as well as drive 16-penny nails with one stroke. The drawback was getting one arm muscled like a lopsided gorilla, so I opted for a 28-ounce hammer. Forty years ago, many houses were framed with 16-ounce hammers, but big hammers are undeniably an improvement.

Which brings us, on this odyssey of hammers, to a word about sledges, mauls, and mallets. After a fashion, they're hammers, too, and for some jobs, nothing else will work. With one blow, croquet-style, a maul can move an entire wall or coax a log along in fractional increments. And no hammer works better than a 3- or 5-pound sledge for setting stakes. In both cases, ordinary framing hammers will bludgeon the wood to splinters. Soft-faced mallets (with striking surfaces of leather, plastic, or wood) are the acceptable tool for smiting a chisel, whether you're chopping teensy dovetails or framing a barn. Barbarians use any old hammer, of course; it's a free country.

The most and least creative uses I ever saw for a hammer were identical in form, if not purpose. One carpenter slipped on a wet plywood roof and was schussing off the edge when he buried his claw hammer in the sheathing like an ice axe. It saved his neck.

Under the heading of misuse, I saw a young framer pulling plywood up to a roof by leaning the stack against the eaves and burying a hatchet in it, sheet by sheet, to haul it up, heedless of the holes he was making. "Gonna be roofed anyway," he said. In this and other ways, he didn't make a fetish of quality and was subsequently fired.

After a few years of framing, it seemed time to move indoors and explore other trades. Burt was a drywall contractor and paroled ex-convict, who had committed some infraction along the lines of homicide but was now back to plying the trade that, he said, had turned him to crime. Two tours in Vietnam teaching the theories of von Clausewitz also had jaded his outlook somewhat. His arms were covered with tattoos such as "Cain," of whom he seemed to be a particular fan. He resembled a musk ox, with murderous eyes, and his face had all the warmth of a hockey mask. But by golly, he knew how to use a drywall hammer.

A surprisingly patient teacher, Burt would answer any question I wasn't afraid to ask. The hatchet blade on drywall hammers, he explained, is an anachronistic holdover from the days of lath-and-plaster, when the laths were chopped, not cut, to the correct length. Nowadays, gypsum wallboard paves over studding in one step, but the hammer still retains its blade. It has no other real function, although I was misinformed by a clerk that it could be sharpened to a razor edge for scoring sheets of drywall. It doesn't work, however; and unless you enjoy an edged weapon heading toward your face on every backswing, better to leave it dull.

A 16-ounce carpenter's hammer works for nailing drywall, but the specially designed drywall hatchet will do a better job. For one thing, the poll is a pronounced mushroom shape, and the dimple it makes is less likely to crack the paper facing of the sheetrock. Also, the head is angled opposite of carpentry hammers, whose polls turn downward about two degrees (toe-in); the drywall hatchet has a toe-out, angling up around seven degrees higher than the perpendicular. This helps when nailing at the top of an eight-foot wall. "You ought to buy one of your own," Burt told me. *Or I'll kill you,* his eyes promised, but that was how he always looked. I bought two, just to be safe.

A word: All the brand names mentioned in this book, such as the AJC, have lots of competitors. Shop around, test a few models and weights of hammers until you find one that your hand prefers. I never met a carpentry hammer I didn't like. Over the years, I've used several Vaughans, a Plumb or two, and many Stanleys. They were all good hammers, and they took a lot of abuse day after day without letting me down. But here follows an unsolicited testimonial: After I retired as a contractor, I discovered the Hart Framer. Without knowing anything of its history, I'd say it was designed by a poet who framed, someone who wanted to design and build a great hammer. So far, I've bought sixteen Harts, and am now down to eight; they never break, but occasionally I give them to other woodworkers as a friendship token. It may not be the best framing hammer in the universe, but then again, it might. Hart also makes a line of finish hammers, nail-pullers, and tape measures, all of which have worked perfectly for me. I should have bought stock in the company.

As a semiretired carpenter, I try to embrace an eclectic approach, incorporating new inventions into my toolchest to join the hoary antiques. But long ago, I sold those relics of my years as a finish contractor: my nail guns. Unless I give them away, I hang on to my hammers, a collection of fifty or so, some of which see daily use. The art of the hammer takes many forms, but its essence is a willingness to keep learning. I am indebted to hundreds of teachers and to those ghostly carpenters of the past, Swanee included, who keep whispering that the only real power is knowledge. And acquiring that takes a lifetime.

He who wonders discovers that this in itself is wonder.

M.C. Escher

SAWHORSE

Nonpareil Trestles

Over two decades have passed, and yet I remember too well the day I learned to build sawhorses. Yesterday isn't quite as vivid. Perhaps it was the blood, a brilliant red puddle beside a dropped saw, and the way blood had spattered the blade, making it look like a dead shark's maw. That old master carpenter, Swanee, had only a high-school education, but he understood the science of mnemonics.

That day, I remember, he built two sawhorses on the spot, stepping around the red ooze and the dead saw, kicking it aside at one point. His thumbprint on those sawhorses made them mine, searing their blueprints into my brain as a reminder that, unlike shit, accidents don't just happen. In the succeeding years, I've never lost more than a drop or two thanks to him. Rest in Peace, Burdell Swanson: I haven't forgotten.

He said something once: "Carpentry may not be a dying art, but it's sure as hell evolving in some questionable directions." That may not be an exact quote, it's been so long. In some ways, Swanee was an anachronism, still sharpening his own handsaws with a three-corner file and set. But he could field-strip a power saw down to the armature for repairs and quickly reassemble it in perfect working condition, slapping the body like a rifle when he was done. He told me a story that led me to believe he had learned nuances of precision from the Japanese during the occupation. He had been a soldier stationed in Japan, and they had treated him as a foreign object at first until they found he could carpenter; then a former Japanese corporal bowed deeply and showed Swanee his own pull-saws and planing tools. In their prewar lives, both men had been woodworkers.

Swanee loved and cared for his tools, and could expound for twenty minutes on the history, idiosyncracies, and quirks of, say, a block plane. He had a few quirks himself. He liked planes, loved chisels, disliked rasps, and abhorred nail guns. He hated certain tools, for reasons that seemed strange to me then but perfectly sane now. He

called nail-gun operators "machine-gun monkeys." (Another vivid mnemonic; when I rent nail guns these days, I always feel vaguely simian.) He accepted the fact that their numbers would increase, but he didn't like it, or any other trend that made knowledge obsolete.

Power saws were good. He said so, calling them a true time-saving advance, even though I was sure they signaled the beginning of the end for handmade sawhorses. A pair of sawhorses are absolutely necessary to cut stock with a handsaw, but with a power saw you can, you think, simply prop the board off the ground on your foot and then bend to cut it off. This works especially well with a worm-drive saw, the fabled West Coast Stud-cutter. Safety inspectors and insurance companies are naturally horrified that anyone would do this. But I've seen a man cut a sheet of plywood in half while bracing one side against his abdomen and catching the far side on his toe: a human sawhorse.

Had Swanee seen it, he would have said: "Think about the last sawhorse you saw; wasn't it all cut to hell?" But people never imagine they'll be injured when they do something perilous. They hope for luck. Another thing he used to say: "Hope springs eternal in the human boob."

Bending over to cut a board would have been unthinkable to him. Likewise metal sawhorse brackets, since all carpenters, Swanee felt, should know how to fabricate their own scaffolds, jigs, and sawhorses from wood. It was one thing to use modern time-savers, but quite another to need them because of deficient knowledge.

At one time you could not walk onto a construction site without finding a pair of new horses. This was long ago, when they were called "carpenter's trestles." It was the custom for many years to build them at the outset of a job and leave them for the owners when done. Today, however, most such sawhorses are weathered singletons laced with hundreds of shallow cuts across the top plank. Occasionally one will appear during the framing of a house, someone's heirloom that is universally used as a portable four–legged stepstool and as seating for the lunch hour. They remain in use until they fall apart, and then their bodies are burned on the scrap pyre. No one throws them away.

Only a few carpenters remember how to make a proper sawhorse. Of the many designs that once were used, most of the bad ones have survived—cumbersome 4 x 4 backs and legs made of studding, attached with impossible cuts and gusseted too deep and poorly. I have a mint-condition carpentry textbook from the ducktail-and-hotrod era, a good book otherwise, that offers just such a clunky style, heedless of the young minds that would fall into error. Worse yet, nothing is said of the sacred two-ness of sawhorses, how they are paired from the beginning and forever, until one of them dies, maybe; not a word about this indivisible plurality that is their other salient point. (I see the blood again, hear the Law: "Sawhorses come in pairs.")

Looking back, I believe it was largely by accident that I learned the craft of carpentry at all. After my sophomore year in the fourteenth grade, I needed a summer job, and that has made all the difference. An old Victorian house in my hometown wore the evidence of restoration underway, apparently being done by one lonely worker. It smelled like employment. Inside, I found an elderly carpenter sidesaddle on a sawhorse, testing a plane blade on his thumb. Two flat pencils were sheathed in his chest pocket beside a watch chain, and he puffed a pipe the color of a bear's nose. He looked, I now regret recalling, intelligent for a tradesman. (I had miles to go. Light-years.)

My stock opening—"Need any help?"—produced no response at all, as if I were invisible. He checked that blade a second time, holding it up to the light, scraping a fingernail along the edge. Then he put it back in the plane and screwed down the cap-iron. As the silence reached its tenth second, I surprised myself.

"I want to learn carpentry." That got him to look up, and I could see piercing blue eyes soffitted neatly under the cornice of his brow. I stammered, "So I . . . do you need a helper?"

To this day I have no idea why I said such a thing. It just jumped out. Looking back, I don't recall that it was even particularly true; I wanted a job, not a career. But lying was effortless for me then, and at nineteen, I had no definition of the phrase "lifetime commitment."

"Depends," he said finally, his first word to me. (Over the next two years, he said it many times. It is a fine answer to most questions.) "What can you do?"

Three seconds to inventory my mechanical skills. "I can drive nails."

He smiled hugely at this, cocking his head. It felt like progress until he asked, "What else? Anything else? Can you set a saw?"

"No."

"Know what an arris is?"

Some kind of wine? A Greek god? One of the Three Musketeers? "No."

"Kerf?"

"No."

"A plinth?"

"No."

"Well, can you at least read a framing square?"

"I guess not." I wanted to break up the rhythm.

"Okay. Think you can do the work?"

Ah, an easy one: "Someday, maybe." Had I said, "Depends," he surely would have thrown me out. Swanee seemed to be nodding, which I thought meant good-bye and an end to this ordeal. "Well, that's good. You're hired. I like honest ignorance."

The next morning, he negotiated wages in two sentences. "You'll work cheap at first. But you'll learn a lot." We worked on the big old house, all three stories, for the summer. Swanee seemed genuinely pleased to have discovered such a blank slate as myself to write upon. I knew less than nothing, which meant that he was paying me much more than I was worth. My hands and mind were strangers in that first month, as we laid the sleepers for a wooden basement floor, jacked up a sagging foundation corner, repaired the leaking gutter that had undermined it, tore out lath strips and plaster walls, drywalled over old bare studs and new insulation, and plastered the joints smooth. On a hot day, we set the porch balustrade to newels Swanee had turned on his lathe.

That old man worked hard, and carefully. It just seemed fast, due to his economy of movement, doing each job once and well. He was a quiet workman, only speaking when he lit his pipe and explained what we were about to do, or to pass on a few words I had never heard before. "These are balusters, and don't call them bannisters. You can call it a railing, but it's a balustrade. And this ball on top of the newel post, it's a finial, not a 'globe'; you remember that word." Then, silently, he'd show me how to find an angle with a T-bevel. We worked, of course, without benefit of a radio, which he dismissed as "the moron's best friend."

My college classes started without me. Every day, Swanee showed me something new, some revelation from his apparently limitless store of building knowledge. I was definitely hooked. Academe could shove their "History of the English Civil Wars 207; Seminal Conflicts of Hardhats vs. Longhairs." That was useless life information. But you could build anything with a hammer and saw and a skill, and I wanted to learn it all.

It took until late September to finish that huge house. I expected to be laid off when it was done, but Swanee gave me a modest raise instead and helped me build a tool carryall. One October morning, we drove up to a large apartment complex under frantic construction, where we had been working a few weeks. The sun had been up a few hours, and workers from other trades were hustling about. Swanee had contracted the finish woodwork, a good-sized job for the two of us, but by then I was able to work on my own to a small extent. Several framing crews hammered and nail-gunned on the unfinished units, running behind and trying to catch up.

"Slower is faster," Swanee said calmly. "They've got those boys doing ten-hour days on this anthill, Saturdays included. That's not good." As always, Swanee filled his pipe as I hauled the tools out of his truck, but this time he frowned when he saw a young carpenter in a straw cowboy hat and cutoff jeans hold a board off the ground on the crook of his instep and kneel to make the cut, scant inches from his tennis shoe. "What the hell?"

I didn't know what was happening. Actually, it looked like a pretty neat trick to me. Swanee struck a wooden match with his thumbnail and lit his pipe. "Set us up inside," he said absently. He walked over, waiting to talk until the saw stopped. He pointed back at our truck, aimed his finger down at the saw and said something, shaking his head. The young cowboy smiled, shrugged, and shook his head, too. Swanee came back, obviously disgusted, and retrieved a stacked pair of sawhorses down from his truck bed. He took them over, plunked them down by the Cut-off Kid's saw, and walked away.

"Damn fool has to learn the hard way," was all he said.

An hour later, as we were nailing trimwork, we heard an urgent yell, and all the noise of construction gradually eased down into silence. Outside, the young carpenter was sitting on the deck, holding his foot as blood spurted through his fingers. In the eerie quiet, the construction supervisor signaled a few others to help lift, and they loaded the injured man into his pickup and roared off to the hospital.

Swanee had come outside with his hammer still in hand. Very deliberately, he dropped it into the loop on his white coveralls. Still stacked, the sawhorses had not been moved from where he had placed them earlier. Swanee stepped carefully around the bloody patches and set the horses a yard apart. He turned to me.

"Think back," he said, jaw muscles tight. He took his pipe out, fumbled with it, and put it back in his pocket. "Now I'm not stupid. Have you ever seen me bend over like that to cut anything, ever?" He held up his hands. They were the hands of an old man, the backs lined with veins and white hair, one gold wedding band. "See? Dammit. Thirty-six years I've been cutting boards, and I still have all ten fingers, don't I? Dammit."

He went back inside and brought out his power saw and hand-tool carryall. Carefully, he put his foot on the bloody saw and yanked its plug free from the power cord, into which he plugged his own saw.

"I'm going to show you how to build sawhorses now. I don't ever want you to make a cut without them. You watch."

He pulled some scrap boards from a nearby pile, set his T-bevel with a framing square, and made a series of quick marks on the boards. He put them on the horses and made his cuts. A few workers were watching him, but they didn't say a thing. He flipped boards and drove nails with a ferocity I had not seen before.

In fifteen minutes he was done. He fished out his pipe and a wooden match. Two spanking new sawhorses stood beside the first pair. All four were identical in height and stance.

"There," Swanee muttered as he lit his pipe. "That didn't take as long as a trip to the Gee-damn hospital." He pointed the pipe stem at me, the end leaking smoke. "Listen up. I want you to remember three things. Sawhorses come in pairs, so don't build just one trestle and try to use it. Every angle on 'em is seventy-five degrees. And the last thing is—hear me well—don't you ever use a power saw when you're groveling in the dirt. Not ever. Stand up like a man, dammit."

And then Swanee did something that fixed the moment forever in my mind: a nice October day, the smell of sawdust all around, the bright patch of red on the earth, his creamy white overalls, the two yellow pencils above the heart that would cease to beat in two years, and the scent of his pipe. He bent to dip his thumb in the blood. He dropped his thumb on the tops of the new horses he had just built and made two even strokes.

"Now they're blooded. Don't get any more on them. They're yours."

There are no accidents whatsoever in the universe.

RAM DASS

TOOLCHEST

The Grown-up's Toybox

Call it contemplation, meditation, or simple examination of the shape of each morning for fifteen minutes after I light the fire. I can spare exactly fifteen minutes a day toward Enlightenment, beginning a few heartbeats before dawn. The family will remain asleep for a little longer, and then we'll all gather for crunchy carbo-loading and caffeine enjanglement. But inspiration comes easier in silence and solitude. For now, I try to still the restless mind, inquiring: *What will it cost me to become a better writer? To understand all the crazy rules of English syntax, to fletch verbs that soar, to obviate untoward verbosity and eschew sexist language; above all, to write well and clearly?*

My stomach growls once. Then, from the misty ether, an answer comes zinging back: *Part with all your tools.* Radical simplification, in other words, a red-hot dose of Zen from the skin-shedding school: Sell, give away or otherwise lose the tools I've honed and oiled for twenty-five years, some of them. It's a logical answer, though. It makes sense. If I spent the time writing that I now spend building, constant practice would inevitably lead to great proficiency.

But give up my tools? Ha. Not the ones in my toolchest, surely. And what about the chest itself?

This is the world of things, a plane where humans frantically gather tangibles into a pile on short-term loan from the universe, less than a century of ownership and we're gone. Yet we form attachments to these things. They are ours.

Ownership, however illusory, carries a price. We must keep track of our things somehow. To do this, and to simultaneously acquire even more things, we have developed a highly evolved brain. Fastest and finest of all computers, the human mind will (or can, anyway) coordinate many functions at once. Like a shepherd tracking each sheep in a flock, those who own a thousand books can probably tell you the very shelf on which to find any given volume. Even in sleep, a professional auto mechanic knows

the location of every wrench and ratchet in her garage. And, after some coffee, I can find all of my thousand or so building tools, in fifteen minutes or less, from an ink pencil for marking wet wood to my second-best froe.

About a hundred of the best, all hand tools, live in my toolchest; the rest radiate out from its center. At one time, carpenters were judged and hired on the basis of their portable tool cabinets, those shiny cabinets that sat on the back of the wagon, as much as by their tools. A good toolchest, solid, serviceable, and maybe imaginative as well, meant a good carpenter, just as the hasp and padlock meant valuable tools within. Those old wooden boxes were heirlooms, built to last for generations: a resumé in wood.

Ideally, such a well–made carpenter's toolchest would be too exquisite to even contemplate using. You should begin with a good hardwood such as oak or maple, cutting precise dovetails at all corners and joints. Now spend a few years tuning the marquetry and parquetry, in oak or koa, with some surgical purpleheart inlay and carved bas-relief images of the sons of toil at work, the whole festooned with locking book-matched drawers and tricky cocobola turnbuttons to hold each tool securely in place—but how could one dare to dirty this toolchest with real working tools? I have seen them in museums and galleries, which is where they belong. They are the orgasmic legacies of a slower, gentler age of wood, before freeways and MTV, when people used to churn butter for excitement.

And they are no longer practical. For one thing, they take a lot of time to build. Maybe there was more time a century ago, more minutes per hour in the sepia age Before Television. Carpenters were expected to take months constructing a good toolbox, and years detailing a full tool cabinet for the shop.

But I say unto you: plywood. Now, don't curl your lip for too long, lest the sneer cramp up. It's all wood, anyway. Marine plywood will last as long and work as well as any form of wood except possibly teak. If what you need is a box that will contain most of your tools in orderly array, something you can heave up in the pickup on a moment's notice, then think plywood, and fir, basswood, even pine for the trays. My own toolchest is simple, functional, inexpensive, and solid. It took two days to build, and it will hold more tools than you can carry at once. I used plywood.

By the time I decided to build one, I had examined quite a few types of personal toolchests. Swanee had made a plain oaken box the size of a small footlocker. His hand tools were racked, stacked, and nested simply, everything accessible and nothing more ornamental than a few brass fittings. Later, I worked for a man whose incredibly detailed and massive toolchest was the handiwork of his great-grandfather. Seven kinds of exotic woods had gone into it; it concealed invisible drawers and stood upright on casters like a teakwood steamer trunk, intricate and complex as the Taj Mahal, a little dollhouse for hand tools as designed by M. C. Escher. No space was wasted, nor any tool crowded. Everything seemed to be spring-loaded, well-considered, double-safed. "Feast your eyes," its owner told everyone. Indeed, it had been built to please the eye foremost. But the workmanship, the sheer time invested, was daunting.

At an auction, I came across an Amish farmer/carpenter's toolchest and was duly impressed. Severely simple, just a trunk with two trays, and yet this chest had a dignity that I had not seen before. It was not crammed to the scuppers with tools, but a few were neatly stacked in its trim compartments, awaiting a hand that would never come. The corners had been rounded off by constant wear and a little whittling. I loved its simplicity, and later designed my own toolchest along those lines.

The most artistic toolchest I ever touched belonged to Carlos Arronde de Luna Socorro y Simón, or thereabouts; he was a coworker of mine in Colorado. We called him Gordo for short, and I never met a happier or more meticulous carpenter, the kind of woodworker whose hands flowed over the workpiece in fluid shaping motions like swallows skimming a gambrel roof. His colorful box featured a little shrine inside the lid, where a tiny ebony figurine of St. Martin of the Poor lived, surrounded by a wealth of beautiful tools. Whether consciously or not, Gordo dropped to his knees when he searched for tools inside the box. The outside of the entire chest was blazoned with scarlet and yellow designs of Aztec origin, and it had eyes painted on one end like a Mexican second-class bus. When he put it up on his truck, the other end gave tailgaters a Mixtec blessing: *¡Adiós!, cabrónes!*

So many choices. But you would not believe how many years I put off building my toolchest, afflicted with a pride that would not let me put my fine old hand tools in a crummy plywood box. They were worthy of better, I thought; maybe a toolchest like the one I saw in a book, built by a German cabinetmaker in the last century. Two years to rough it out, another year to fine-tune tool placement for the balance . . .

At the time, I lived by the sea. My tools were in danger of turning brown. When your high ideals are at odds with your environment, examine them both. In 1983, having practiced carpentry for a dozen years, I finally built a substantial chest to replace a battered carryall that had become a barely portable tool catchall. It took a day to build. This new chest was not portable by fewer than two human beings, unless I took all the tools out. But that was my intention.

First, I sketched out the basic plan on a piece of drafting paper. It would be a closed box, with a gasket to keep tools safe from the corrosive maritime air of our little coastal town; it would be less than a cabinet, more than an open box, a practical chest for tools in the shape of a wooden trunk. I thought again of that old Shaker toolchest, with only a few coats of brown paint for protection.

To build one exactly like mine, buy two sheets of good ¾-inch plywood, A/C. The basic dimensions are 16 inches high overall, 32 inches long, and 18 inches wide; make the lid 4 inches deep, which is deep enough when closed to house a score of tools that won't rub against those in the top tray.

There are two stacking trays inside the main well, and both trays are divided into sections to keep tools in their separate categories. Otherwise, they'd flow together like sand. The bottom tray is a foot wide, and it rests on the first level of partitions that compartmentalize the bottom of the toolchest. This narrow first tray creates a small pocket at the rear of the toolchest, just tall enough to fit a few handsaws. The top tray sits flush with the top of the chest, except for the side handles, which stick up a little. But there is ample clearance for all the tools in the lid.

So—I made a box in the above dimensions, initially with no openings. The side joins were simple but adequate half-blind joints: a rabbet in the front and back panels to accept the side panels. The bottom and top panels were fitted a little more intricately, blind-rabbeted, since they would receive the most stress. But no dovetail joints anywhere; it takes thirty minutes to make the average dovetail.

I used waterproof glue and carefully placed brass screws to assemble the body, clamping it nine different ways overnight. (Do not use nails. Really.) In the morning, after coffee, I cut 4 inches from the long side of the top by turning the box sideways and running it through a table saw in four long, careful passes. (No nails, remember? Keep your screws off the 4-inch mark, too. The thinner the kerf, the better the fit, by the way.) On the fourth cut, shim and scab the other three sides so the blade won't bind and take a bite out of your pretty toolbox. Voilà, lid and body, to which I screwed a brass piano hinge on one side to join them. Finally, I made the trays.

Arranging the tools took another day of study. To qualify for a spot in the lid, a tool had to be fairly small, flat, and light—no jackplanes or bitbraces—and used fairly often. All of my Yankee screwdrivers, for instance, found a place thereon, as did both steel bevel squares, the framing square, a short level, a compass, and a few other favorites. I used fir for the racks and oak for the turnbuttons: those little tabs of wood that turn to secure tools in their racks, recesses, sockets, crannies, niches, and pigeonholes. Each tool must remain securely in place when the lid is down, or what's the point? Even a minor rattle becomes intolerable after a while.

Fully loaded, it probably weighs over two hundred pounds. I no longer take it out of my shop; the lid stays open for months at a time, always ready for me to snatch a tool and put it to work. It's no work of art, but nothing much short of death or theft could pry it away from me. These tools, and certainly the toolchest that houses them, are mine.

Morning begins. Any other messages from the old oceanic superconscious? Tiger stumbles in, a walking argument for simplicity, following the path of least resistance to his food dish. He plunges his face into the dry soymeat, eyes closed. I hear a few stirrings of life from the other room.

First, coffee. Then, it's time to write this all down, as best as I can in my own clumsy way, heavy on the style, easy on the numbers. Excelsior.

The fewer our wants,

the nearer we resemble the gods.

SOCRATES

STRINGLINE

Getting Straight

A few well-placed lattice screens block the view of our house from passing cars, because I don't recall moving to the country to be observed in my natural habitat. What was once a highway, before the Oakland earthquake, is rapidly becoming a freeway. The screens form a psychological barrier, something to hide our lives behind. But until the next oil embargo thins out the traffic, I'd prefer something high, opaque, and sound-reflecting, with a simple "Private—Keep Out" sign in blazing fluorescent letters.

Placement is the first step in building a fence. Two stakes fifty feet apart define its span—I put them in yesterday—and connecting those stakes with a dryline will be the second step. Early in the morning, I gather up my walnut string butterfly and slide an awl into the copper-lined hole drilled through its axis. Starting from the first stake, I pay out neon-yellow string from the tumbling spool.

If there's one secret to building well, it's being able to establish a perfectly straight line and follow it. To be truthful, a string isn't absolutely necessary to align the posts and rails, or to keep the top boards straight. You can eyeball everything, as a former neighbor of mine did with his fence. "Not very pretty, but it's functional," he said. All it needed to make it pretty, in my opinion, was to wait until nightfall, douse it in gasoline, and toss a match.

With real interest, our cat watches the line tighten, perhaps contemplating loftier matters than the path of proposed fence-post holes. He is not the brightest cat you will ever meet, but Nature has given him sufficient intellect to rest, eat, meditate, and generally spend the day sponging off the big monkey. In return for various blandishments and delicious Mouse Helper, he makes a soothing nasal sputter and allows us to overfeed him. This is a perfectly satisfactory arrangement.

But when he decides to twang the string with a claw, I shoo him off: "Take a hike,

Tiger." He looks up, his expression conveying about ten thousand years of apathy for the human riddle, and waddles off.

Depussified, Ozymandias resumes building. In construction, the idea of perpetual stasis is an order of folly higher than perpetual motion. I'm thinking of a fence I built ten years ago; no wiggles, strong as ever, it still runs basically true after years of midget ninjas kicking and scaling it. But entropy and the elements have drawn a vague curve in its run, as if it were trying to match the curvature of the earth. It's a comfort to know that any straight lines I create during my lifetime will be altered by the passage of time into planet-friendly arcs. Joseph Campbell called the straight line "a circle of infinite radius," which explains just a hell of a lot.

That said, I like my lines straight. It's a peculiarity of structures that their lines must not meander, so they'll tend to last longer: the real point of civilization.

The dryline illuminated one mystery of carpentry for me. It seemed like magic, some inner measure or instinct, that enabled a carpenter to build in orderly straight lines from which sturdiness derived. Nope; just string.

Over time, I began to see how complex a tool it was: Yellow string, for instance, gives better visibility against a dark background than white string. Nylon is ideal for dryline work, taut as a bowstring at full stretch and more resistant to abrasion. Cotton has advantages over nylon string; it works slightly better for chalklines, and much better on a plumb bob because it won't stretch or rebound.

Most drylines are wound around H-shaped reels made of anything from plywood to rosewood. Mine is oiled walnut. A long time ago, I made a butterfly out of thick acrylic, thinking in terms of utility and permanence. This was a great idea, and it was absolutely unbreakable up to the day it vanished.

It doesn't take long to dig all the holes. I set the end posts and drive a small 6-penny siding nail, two inches long and slim-shanked, into the top of each post. The nails are driven at a slant into the top outside arris—a useful term, meaning the 90-degree angle formed at the intersection of two planes. The top of a 4 x 4 post has four corners, and four arrises at those corners. The string should stand away from the fence about a quarter-inch. Use a small block of wood under the string, right next to the nail. This offset—the distance from string to the wood—is called a reveal (or back-off), and it keeps the string from being pushed out of true by the intervening posts as I set them.

There's a permanent loop tied at the end; this goes over the first nail. At the other post, I make a loop over my index finger, twirl my finger in a circle five or six times, and slip the loop over the nailhead. Pulling the string back the way it came makes it taut, and a gentle yank in the opposite direction lets the loop cinch itself into a neat little hangman's noose. To release, just tug the loose end in the direction of the taut string. It always works, but twenty years have not dimmed my amazement.

I love string. If an old beam sags like the belly of an elephant, a dryline tells me how far to jack it up. If I want to check how badly warped a large surface is, such as a floor, ceiling, or wall, I stretch a string across it, and the hollows appear like magic. Affixed to a plumb bob, string will always point toward the center of the earth, ensuring that a porch overhang is perfectly vertical to the deck beneath, or that a proposed hole in a roof will be directly above a vent pipe.

Nails aren't the only thing that will hold the ends of a stringline. Masons use a pair of ancillary tools called "dogbones," which are wooden blocks that hold both ends of a string. When you can't drive a nail because of the stock, as with bricks, or when you want a sliding line for lap siding, dogbones will keep tension on a horizontal line and allow it to be moved up in increments, without slacking. Some masonry supply houses give pairs away free, with advertising printed on the wood.

The chalkbox is a canny modification of the basic principle: an encased reel holding fifty feet of string dusted with colored chalk. Stretch it from point to point, snap the line, and a ghost of its path remains. On long runs, a finger holds it down in the middle and both sides are snapped. Thus, a bow in a rafter can be marked and cut away, or courses of roof shingles kept from wandering.

For general carpentry, I like blue chalk because I can see it better. Red is less visible on cedar siding, but disappears nicely under stain; white chalk works best on green cement and roofing felt. To avoid mixed pastels, I keep three chalklines of varying vintage, a small extravagance. In my early poverty days, I used one chalkbox for everything, even a make-do plumb bob, but not after finding a brass bob of exceeding antiquity, which cured me of pointless thrift.

There are always compromises in the name of economy, but one of them doesn't have to be string. It's cheap. A few years back, fencing with an uncarpenter friend who had no panache to speak of in mechanical matters, I used his . . . ball of kite string . . . under protest. It worked, of course, but one might as well use old bits of knotted twine or braided nose hairs. It wasn't the Right Tool, I insisted. But he couldn't see the distinction. Maybe there isn't any.

After setting all the posts, a small matter of six hours, it's time to give the string a final check. TV beer commercials show barn raisings with laughing male-bonders and nail-pounders, all guzzling suds in a work party for Amish yuppies; but the best moments of carpentry take place in solitude and silence, standing atop a ladder and sighting down a string to check the tops and sides of the posts. The fence will be true: in this case, it means the truth of the line, which the posts follow without deviation.

An executive luxury chariot rockets past at seventy-five m.p.h., spots a hick on a ladder, and honks. In the country, people in cars honk at people on foot. I smile and wave: Take a last good look, outlander. Tomorrow I'll install the rails and sheathe them with seven-foot cedar boards, both guided by stringlines; and long after I'm gone, my fence will make a high, sun-warmed perch upon which cats can soak up leisure all day and conduct their amours under the moon.

Example is more efficacious than precept.

SAMUEL JOHNSON

25'
STANLEY
POWERLOCK
II
GESETZLICH GESCHÜTZT
Watermeyer, Hildesheim

TAPE RULE

Measure for Measure

In the beginning was the stick. It aided the process of ambulating on two legs, served as a club for settling disputes, and literally grew on trees. It had balance, reassuring the hand and mind. Best of all, it stayed approximately the same length from moment to moment. This was useful for communication. Tell them around a new fire that the home cave was one stick high, and they knew exactly how high; they all had sticks themselves, for clubbing things. Pretty soon, everyone was leaning on their sticks, laughing and modulating grunts.

In time, the old stick upgraded to a stick-with-marks, decorated with little notches at appropriate intervals and crusty brown stains of special significance. By then, it served Man-Who-Thinks-About-Sticks as a diary, a calendar, and a measuring device, but only for measurements up to one stick long. In those dark eons before color television, he pondered his stick for hours. His flowering intellect wondered: To keep it from rolling, could it be shaved flat? Probably. (Memo: invent knife.) It might not work as well for settling who got the haunch, but it would be superb for determining linear distance, even in multiples of stick, once he figured out how to do it. At the limits of the firelight, a huge snake looped itself down from a hanging tree branch. Ugh. Now, if you could make a lo-o-ong stick coil, somehow . . .

Concepts were fuzzy, but they kept coming. Under the ancient moonlight, Man-Who-etc. dreamed of counting, music, agriculture, knots, measurements, shelter, rituals, hot tubs, and, maybe someday, a stick that rolled up into a tight coil. Heady stuff, this thinking about sticks. From the cave, Morning Person inquired when he was coming to bed.

Eons later: "Soon, honey. Before midnight," I tell my wife. "I need to work on those cabinets for an hour or so." Joy knows that I like to spend wee hours in my shop (our shop during daylight hours), freefalling into a coma of pure thought. Some of my most intense and productive thinking takes place late at night, when the world is silent. Surrounded by tools and dozens of projects-in-progress, a very old part of my brain relaxes. I know where I stand in the universe. And exactly how long a stick is, whatever that means.

Brightly lit, warmed by the attached greenhouse on the south, the air perfumed with cedar sawdust and the scent of oily tools, this shop is where I play with wood. All around the workbench, disorganization reigns. It has been ages since the shop was cleaned. As usual, I put it off again.

Behold the board. It just sits there, a twelve-foot length of pine heartwood, waiting for me to shape it. With that crazy grain pattern, it cries out to be made into another pair of book-matched cabinet doors. But a power saw would shatter my delicate mood, not to mention waking everybody up. All I can do is mark off sections for tomorrow, using my coiled-up measuring stick.

I reach behind to a spot just over my sacrum where I keep the tape measure on my tool belt, but my hand grasps nothing as the motion completes itself. I seldom wear a carpenter's belt in my shop. Part of that has to do with the plethora of flat surfaces on which to lay tape measures and pencils. When they walk off or merge with the background of tools stacked upon tools—the main drawback of the flat-surface organization method—I find myself feeling where the tape holster should be, right over my coccyx. It is an unconscious habit, buttressed by ten thousand similar reaches during my working years.

I look around. So where is my tape? Somewhere in the mess.

When I do wear my belt, the routine is always the same: draw the tape case in one smooth motion, extend the tape with both hands (or hook it on the board and pull the case away), read the measurement, look back to make sure the hook is still where I want it, check the measurement again, retract the tape (one thumb braking its return, for the sake of the hook rivets and mainspring), and back it falls, snug in the pouch. My hands are working the square and pencil even before the spring unwinds itself with an eerie buzz at the small of my back.

There, on the workbench, is my steel tape. It is a 25-foot Stanley that has been with me for donkey's years—not the same one, but the same brand and model of self-retracting tape measure. Some carpenters may prefer other models, but woodworkers tend to be brand-loyal. Back a few decades, it was called the push-pull rule, even when it was spring-loaded, as all are now. I don't remember if the first one I bought had a metal case, but this one is made of a tough plastic coated with silvery paint. It will not dent, and I have never had a case break. The moving parts sometimes failed under hard and daily use, but in the present day they are made of sterner stuff.

Many manufacturers offer 25-foot tapes with cases in neon fluorescent colors of pink, orange, yellow, and green; these are not only pretty but highly visible. I am going to buy Joy one, any day now.

Another improvement is the three big rivets holding the hook at the end. When I bought my first tape over twenty years ago, its hook had two little rivets, which often ripped out when the tape was reeling back and the hook caught on a board. Three rivets keep the hook well secured. If you drop it and it lands on the hook, it can throw the measurement off until you bend the hook back to shape with needle-nosed pliers.

It had to be explained to me, because I would not have noticed on my own, but this hook at the end slides along the rivets, exactly 1⁄32nd of an inch: the width of the hook itself. This keeps the measurement accurate whether I am pulling or pushing against the end of the tape. The tape is slightly concave in cross-section to make it stiff. The arc is a full seven feet, meaning that I can extend it out that far horizontally from its case before it bends. This stiffness is handy for measuring vertical distances: Pull out eight feet of tape from the case, grab it at the four-foot mark, and bend the tape into an arched loop with the numbers facing you. Put the hook on the floor, guiding the tape with your free hand. Now, with the case in the other hand, you can crawl the tape up the side of the vertical surface by pushing the case toward it and read the mark easily without going up on tiptoe.

Under heavy use, the first ten feet of tape always wears out before the last nearly pristine fifteen. But the size and heft of this measure exactly suits my hand. I cannot imagine using a shorter tape, although Joy can, and so she does. Her 16-foot tape measure feels like a toy in my palm, and anyway, a smaller tape case would rattle around in the tape holster.

When I met him, Swanee was just making the switch from the wooden 6-foot zigzag rule, once the standard carpenter's measuring tool. He still kept a zigzag in his toolchest, and he could flick it out to its full length with a snap of his wrist. I tried, but it seemed a very primitive and clumsy device to me. He assured me it was quite accurate, although it took years of practice to master. I have never bothered. For sheer utility, the steel tape is head, neck, shoulders, trunk, and groin above the zigzag.

He also had an old boxwood folding rule, called a "one-foot four-fold" for its extended length and number of sections. At the very end is an integral brass caliper that adds another two inches. This antique but clever tool is perfect for measuring boards to be glued up into cabinet panels. I still use it for just that. Before he died, Burdell Swanson gave it to me, along with many other gifts.

He was a stickler for exact measurements, the hallmark of a master carpenter. "Measure twice—in your case, three times, if you have to—and cut once," he advised me on my first day. "Learn to read that tape from any angle, and get your number to within a sixteenth of an inch." When my cuts got sloppy, he called out measurements in tinier increments: "Make it a strong five-sixteenths," meaning I should err on the side of too long, say 11⁄32nds of an inch; but that fraction is a headache to figure out. The adjectives strong and shy are easier to translate. Lay the tape on the board and mark a hair beyond or before the ultimate sixteenth of an inch. You'll never get it that accurate, as Swanee well knew, but the attempt teaches you to be exact, which is what he was trying to teach me.

By the way, make a caret with the pencil, instead of a short line. A little arrowhead is easier to see and provides an exact reference point. Some people always point the caret toward themselves, marking from the top edge of the tape; some point it up, using the bottom edge. I make carets on the top edge when measuring from the left, and the bottom edge when the hook is to my right. Try it and you'll see why.

Dimensions are critical in construction. I have been on sites where work stopped altogether when a tape broke.

To be truthful, I was the only person present. In contracting, lost time is lost money; I don't build professionally anymore, but there are always three or four 25-foot tapes in my kit at any given moment. Or five or six.

Keeping them company are 50-foot and 100-foot flexible steel rules with circular leather cases. Both are old Lufkins, and the longer one still looks nearly new after forty years of service to two owners. These are not spring-loaded; they must be rewound with a little handle that pops out from the center of the case. But it is kind of fun, like fishing for dimensions. The sturdy hook at the end of these long tapes does not slide because they are a pull-only type of measure; you set the hook over a nail or board edge and walk away, keeping tension on the tape.

Recently, I acquired a Hart 100-foot self-retracting tape with an open case that is self-cleaning. Originally a logger and lumber-scaler's tool, it's the best tape for wall construction that I've ever used; wish I'd had it twenty years ago. Before that, I used an old Lufkin and was perfectly happy. I traded a brand-new Lufkin for mine, since it had been in the original box since 1952 and had a real old-timey feel to it.

One of the first things to learn about any tape measure is how to read it upside down. Do it carefully. When in haste, it is astoundingly easy to mistake 39 inches for 36 and thereby make your cut 3 inches short; upon discovery, you will rip the welkin with expletives, to no avail.

Another technique that requires mental attention: When measuring a framed wall diagonally to ensure squareness (the diagonal measurements should be equal, or the wall is not square), I often "burn" an inch. This simply means holding the tape to the corner at the one-inch mark, for extreme accuracy. But if you forget to add back that deducted inch, your workpiece will be an annoying inch short. More oaths ensue.

Time may be relative, stretching and contracting, but measurements are not. They stay the same, no matter how many times you measure. The hoary advice—measure twice and cut once—still makes sense. You can catch a lot of errors that way.

Two in the morning already? Time to crawl under the skins. My board is marked out for tomorrow's cuts. And just to be safe, I'll check the measurements a third time, after breakfast.

I know of no way of judging the future but by the past.

PATRICK HENRY

Tyvek
Housewrap
Tyvek
Housewrap
Tyvek
Housewrap

LADDER

Getting High

It was one of those old movies in which hordes of Saxons were storming the fortifications of a castle, with slaughter and havoc enough to delight the heart of any child. I was seven, old enough to appreciate the tactics of both sides. The assaulting force put long siege ladders against the embrasures of the ramparts; a bad idea, because the clever defenders waited until the ladders were loaded with Saxons before pushing the tops away. Stunt people and the arrow-impaired rained down.

That night I found myself weighing the unlikely odds of survival, and had the first of a series of dreams: high aloft, over the deafening silence of battle, I was falling backward on a siege ladder, an exhilarating sensation that always woke me up before I hit. Alternately, never on the same night, I was one of the defenders, tipping ladders off the bastion with a long forked stick while catching crossbow quarrels in my left ear. For years, the dream recurred monthly; when I entered puberty, aged by years of nocturnal combat, they stopped on their own.

The point is this: As an adult in this plane of reality, I am always careful with ladders because I know how easily they can be toppled and exactly what it feels like to ride one down.

Humanity was not meant to rise off the ground unaided, but when nothing but altitude will do, ladders make it possible: picking fruit, painting eaves, installing chimneys, cleaning roof gutters, even rescuing treed cats (although, as firefighters say, when was the last time you saw a cat skeleton in a tree?). No homeowner should be without a dependable ladder or two. Just one little caveat: Used improperly, ladders can injure, maim, kill, or scare you badly.

In the world of construction, ladders are considered indispensable but dangerous. No carpenter is without ladder stories, experiential or overheard, usually with a valuable lesson inside. I have been on job sites where ladders were set up with a bit of ritual, at the very least a preclimb check.

Some went beyond that; I knew a roofer who took a full ten minutes to set up his extension ladder against a two-story building. After setting both flexible treaded feet securely on even ground, he drove a stake beside each leg and immobilized them with two big C-clamps. Then he went up, one rung at a time. When he reached the top, he screwed a pair of small eyebolts into the fascia and used a cord to tie off the ladder top to the roof.

I asked what made him such a yellow-belly, although I used the word "cautious." He flashed a look I won't attempt to describe. Had I, smartass, had I ever been on a ladder that decided to slide off a roof, he asked? Actually, no, I replied, but—

He had. Fun once, he said, but not twice. You got your choice, he said: dazzling pain, beige hospital food, soap operas, itchy casts, stupid questions, doped for months on painkillers, and really sorry your Mom had careless children, or the reasonable alternative: Always Secure an Extension Ladder, Top and Bottom.

Whoa, you might say. That's a lot of work, not to mention the time wasted. After all, ladders must be inherently safe, or people would just sue the hind legs off somebody. You are correct if your point is that everyone has to die of something. According to Harper's Index, one-fifth of the cost of any given ladder pays for defending lawsuits against the manufacturers. Few are successful. A jury can't help thinking a plaintiff needs breathing lessons if he or she does not notice the many warning stickers plastered all over new ladders. If the ladder is second-hand, a reasonable and prudent person also would have to wonder why the original owner sold it to a scofflaw of gravity. Case dismissed.

If I seem overly concerned about ladder safety, it is because I have witnessed a few spectacular falls, all of them caused by lack of precautions. Here's another one: A few years ago, my brother leaned his aluminum extension ladder against his roof to clean the gutters. Since it wasn't a real construction task, he didn't secure the top or bottom, a net savings of five minutes. Against my advice, he saved ten more seconds by neglecting to ensure that the sliding sections were locked in place. They were not. When it collapsed, he fell to earth like a sack of wheat and broke his wrist. I drove him to the hospital, listening to recriminations and vows about ladders all the way.

He was more fortunate than one young carpenter I worked with in 1977. I did not see the incident, but I thought of the roofer when witnesses told me what had happened: He had almost reached the top of his extension ladder when the right foot of the ladder punched through into a gopher tunnel. The ladder began sliding to starboard. At eighty degrees of tilt, its rider looked around wildly—what was happening? By forty-five degrees he had it all figured out, including destination. As the concrete rushed up to greet him, he began to yell a compound word, completing it just as his wife was kissing his forehead in the recovery room. Or so his friends claimed.

I should not forget to mention the unluckiest class of all, those people who have accidentally touched aluminum ladders to power wires. It is easy to forget the top of a ladder while you are moving the bottom; there are anecdotes that would make it memorable, but they would put you off barbecue. Let this suffice: when touching a ladder, Always Look for Electrical Hazards.

Using electric tools on an aluminum ladder, or on a wooden ladder in wet weather, are two more avoidable examples. I suggest using cordless electric tools or a non-conductive ladder made of fiberglass. Neither is cheap, but the cost of purchase is offset by the loss of earning potential that usually follows electrocution.

Over the long run, extension ladders and stepladders are equally dangerous. A stepladder cannot be easily secured in place; it is stable while freestanding, but its stability decreases with your every step up. The forelegs should *always* be fully extended to the final click of the hinge, and you should test the first step with all your weight before ascending. Do not grasp the top and jiggle your body to stance the legs on a wooden ladder; it loosens the step braces. Note how the top tread of a new stepladder is festooned with multiple warnings: NOT A STEP. Read and heed, reader. After the sticker wears off, this injunction is still enforced by gravity, a law unto itself. The only safe time to stand on the top step is when hanging yourself; you might fall and break your legs before you get the noose tied, thereby saving your life.

One would think that no sane being would ever stand on the folding paint-tray shelf found on some stepladders; but yes, they do. This is evolution at work. We should not interfere with these people and their destiny.

When you lean an old wooden ladder against a house, a rope running from the bottom rung to the nearest point on the house will keep it from sliding straight out at the bottom. If you secure the top as well, there's no place it can go. But why bother? Omit both stabilizers, and you have invented a potentially lethal way to go up slowly and come down quickly. All you wanted to do in the first place was to get up near that wasp nest and break it loose with your bare hands. Now, mangled and broken, you'll have to drive to the hospital in your old pickup, after you check the gas level with a book of matches.

Let's move on, because surely you would not commit these other common (and often final) misdeeds, such as (a) leaning a folded stepladder against a wall, (b) holding a paint can in one hand and a brush in the other while working on a ladder, (c) mounting any ladder in wet or freezing weather, (d) reaching or leaning far out instead of moving the ladder, and (e) using a ladder on an uneven surface, especially a staircase. You can buy special extensions for the legs that will make it possible, and certain types of multiple configuration ladders are designed to be set up on stairs, but an ordinary stepladder is not. You will probably fall, and it will hurt.

Ladders are rated by type, with varying duty ratings: Type IA holds 300 pounds, Type I can hold 250 pounds, Type II will support only 225 pounds, and Type III a mere 200 pounds. Duty rating refers to total load. A 200-pound worker carrying 25 pounds of tools and/or materials is a 225-pound net load. Obviously this person should use Type I or even Type IA.

Ladders are commonly made of wood, aluminum, or fiberglass. Having used all three, I prefer fiberglass stepladders. Among other reasons, they do not conduct electricity; they also require less maintenance than their wooden cousins, on which the step-brace nuts must be periodically checked for tightness, and all parts brushed with linseed oil once a year. Never paint any wooden ladder, because paint hides cracks, the first sign of structural weakness. The second sign is failure, always while in use.

My trusty aluminum extension ladder cost less than a third of its fiberglass cousin, and it requires only periodic waxing along the channel that connects both sections and an occasional drop of machine oil on the pulley. Personally, I like the light weight of aluminum, but electricians wisely refuse to climb anything but expensive fiberglass ladders. Someday I'm going to buy a multiposition ladder, which can be folded and locked into a stepladder, extended ladder, or scaffolding trestle. They will reach jobs other ladders cannot. If you have one, oil the hinges now and then, and check the locking mechanisms before using it.

There is not much point in saying this because no one will believe it: an extension ladder is not a work platform or scaffold. It is merely a device on which you can climb to a stable work surface. Check with OSHA if you don't believe me. But few can afford the time and money it takes to build or rent scaffolding, especially for small jobs. No matter what I say, many will continue to stand on a ladder while working. Now consider how seldom you meet anyone with three arms, and you will see another problem: people tend to work on ladders with both hands occupied. Loss of balance accounts for a high number of injuries and deaths every year.

One type of ladder we have not mentioned yet is the stepladder specifically designed for picking fruit, having three legs instead of the customary four. This third leg allows a fruit picker to position the ladder deep inside the foliage. All is well until someone takes it out of the orchard and tries to use it for rescuing a treed cat, as I did once in my green years. What happened next taught a valuable lesson, which I pass on to you for less than I paid: Think Before You Mount a Ladder.

No one reaches a high position without daring.

SYRUS

CRAFTSMAN

TRANSIT

Shooting the Invisible

If you work on your own vehicles, you should try to avoid doing so on gravel. Inevitably, you will drop some tiny part: for example, a special integral nut-and-washer for a foreign pickup carburetor, of a type unavailable in hardware stores. It will tinkle down through the engine and disappear into a field of small dirty rocks. Attempts to find it will only bury it deeper. You will never see or hear from it again.

This has happened to me more than once, and once was too often. I have resolved never again to lift a pickup hood unless I am standing on concrete. Although my intended pad between house and garage will wind up costing many hundreds of dollars, its entire function, as far as I'm concerned, is to make possible the retrieval of small washers and nuts costing about ten cents. I'm delighted with this economy.

But few things are more apt to unsettle the mind than the knowledge that ten cubic yards of concrete will arrive on the morrow. Ready or not, a giant truck will appear just after coffee, and it will dump enough gray slush to fill twenty bathtubs. Via a chemical process known as hydration, whereby a liquid becomes a solid in the blink of an eye, this glop will harden fast and irreversibly. Errors literally will be set in stone.

In theory, my forms will contain and define the shape, and I've never had a form blow out yet. But I saw it happen once: a disaster. So I'm a bit edgy.

In the words of the poet Joseph Walsh: "Sometimes I sit and stare at the rain . . . ." And into every life, a little of it must fall. Just not tomorrow, that's all I ask. Ideally, this concrete pad should shed rain. I want the sides nearest the house and garage to be slightly higher than the end-points where I've buried perforated drain pipe to escort the water away, a difference (or fall) of about two inches. Nothing could be simpler to arrange; I'll just shoot the forms with my level-transit.

"Dad, where did you learn how to do this?" Serenity is watching me sight through the scope while my wife, Joy, holds the control rod. Well, now. I have been waiting nine years for our nine-year-old daughter to notice how clever her father is, or at least how much he knows about this sliver of the sum of human knowledge. Her question is timed perfectly, in part because it provides a segue to a flashback. "Well, sweetheart," I begin . . .

It was on a blood-harrowing cold morning that I first made the acquaintance of the transit. Burdell Swanson, my first and finest boss, had let me watch as he set one up on a job site. But I was not encouraged to stand around sucking up information instead of doing something useful, like standing where I was told and holding a stick with numbers on it. Just doing that correctly was tricky. "Keep still! Hold it away from you!" Swanee shouted, because in those days I didn't have an eye for plumb and I was freezing to death standing in one spot. "No, not that much! Come on, wake up, son."

The site began as a little hillock facing southeast, with a spectacular view of ordinary foothills. There was a lot of tall grass and a few birds circling in the autumn sky, but nothing indicated that the structure of a house would be erected there in less than a month. I had expected something to mark the site: string, stakes, batter boards, even a line of chalk applied directly on the grass, anything. This was to be my first house built from scratch.

Swanee set up his transit, leveled the scope with the four knobs, and sent me out to hold a stick with inches and feet marked on it, standing at each corner of the "house," including two ells and a garage. (An ell is a room or extension of a room built at right angles to the main part of a house; I learned the word that day.) Moving me around from point to point on the dirt like a giant lawn jockey, Swanee sighted through the scope and wrote down numbers.

When the excavating was done and the foundation poured, we repeated the process. Swanee wanted to know if the foundation was level, and aside from a half-inch high or low at a few corners, it was. At least, I thought so. "Well, this is one lousy foundation," Swanee growled. He had higher standards than mine.

By osmosis, I divined this much: The telescope is placed atop a tripod whose legs are firmly anchored in the earth, and this apparatus is made level by turning four adjusting screws. By sighting through the scope, one can read the numbers on a measuring stick held (perfectly still and vertical) by another person. Thus, the relative height of any point can be determined over great distances because the transit is always level no matter which way the scope points. Later, I confirmed my observations by using a level-transit on my own.

For example, when sighting through the telescope this morning, I see the crosshairs resting on 30 inches. I jot down this number in a notebook as my wife moves the rod over to another location, a bit west and somewhat closer. "Uh, kind of hold it away from you a bit, honey. Nuh. Sort of move . . . Okay; wait, not that much, I can't see the damn thing. Toward you more, then. Look, straight up and down is what I want . . . What? I'm not yelling." But Ren nods, siding with her mom. "Was I? Sorry. Excuse me." As stated, I'm a bit nervous about tomorrow. A raised voice? Oh, they'll never know the verbal abuse I took, learning about this one tool.

I learned a special word that day. "Don't want this tool to be OOL," my employer advised, letting me hear his acronym for Out-of-Level as he turned the leveling screws two at a time, making the scope level on the tripod base. It had a slimy, slippery sound. Drop it, bang it, or loan it to a fool: that was how easily a level-transit could slip into a state of untrustworthiness he called OOL.

I traverse the telescope until I can see the rod, turn the focusing knob until the number is no longer blurry, and I find, there it is, 29 inches. The second spot, therefore, must be one inch higher than the first. If that doesn't make sense, close your eyes and visualize it. The rod—in this case an old tape measure affixed to a flat stick, the hook attached at the bottom—went up an inch, and the mark for 29 inches was below the mark for 30 inches. It comes up to the crosshairs. You see?

"Is that a telescope?" Ren asks. Yes and no. I explain that a telescope is one part of its assembly, and that lots of people call the entire tool a transit, but some call it a builder's level or a dumpy level, and others a level-transit, and a few would say it's a sight level . . .

"Actually, that's more than I wanted to know," Ren says. In that case, I won't tell her that a transit, strictly speaking, is one that will pivot up and down in a vertical plane, so that many so-called "transits" are misnamed. Or that the control rod (also called the leveling rod) is dubbed a Philadelphia stick in some parts of the world. She won't care.

For the purposes of my pad, I'm tickled about that inch of fall. If we were shooting a house site, we would have a very level spot, practically perfect. However, if we were shooting a foundation and one corner turned out to be an entire inch low, this would not be so good. The sill plate would have to be shimmed up with shims made of steel; wood shims would compress too much. When you put a transit in the center of a new foundation, you can shoot every corner, and by using shims, make them utterly level with respect to each other and all the midpoints.

The purpose of the level transit cannot be said any better than this: Let it be level. In building, this means creating a decked platform of wood and concrete upon which you will erect walls, a roof to rest on those, and the surface you call the floor will be so level that, given a seamless floor, you could drop a marble anywhere and it would not roll.

There are only a few ways to go wrong, but one-hundred percent of them will put you entirely out of level. If you fail to anchor the tripod firmly, any of its three legs can slide out, rendering the transit OOL. You also may forget to turn the telescope ninety degrees after leveling it along one axis. But don't.

If you work carefully, a transit can be used as I'm using it, to ensure that concrete forms will be low at one point and high at others, or to lay out the corners of a building, locate the height of grade stakes for footings, run a straight line for fence posts and roads, even to plumb up a tilting wall. It's versatile.

A basic transit will cost about three hundred dollars to buy, or in the neighborhood of twenty dollars a day to rent. If you have piles of money you are tired of tripping over, you might invest a thousand dollars or more in a laser transit, the choice of professional surveyors. It will never be anything but dead accurate because electronic alarms will announce when the base is out of level.

So far, a laser transit is the last word in this technology. One type whirls around like a top, making a solid line of light inside a building; this is great for installing a dropped ceiling, obviously. Once in my life I used one, and only a supreme effort of will coupled with a lack of funds forestalled its immediate purchase. But it was a close thing.

In one of those odd bursts of synchronicity that make life on earth such an inkblot test, the man who gave me my current transit had only one eye, the result of a traffic accident. He called it his "transit eye," implying that it had special carpenterial powers; he was a tough old bird named Morris Salters, eighty-one years old, and as he put it, "Every single one of 'em was educational." When I met him, he was building an urban house to rent out for the income, and I helped him for a few weeks. Among other things, he taught me how to make illegal but semisafe quick scaffolding. This dumpy level was part of my wages. It's as old as I am, but a lot more precise.

After some tinkering and stake replacement, the forms look pretty good. I thank Joy for her help, apologize for my weird mood, and begin to take down my transit as Ren comes up, holding something in her palm. "Dad. Look what I found."

Something shiny caught her eye, and there it was: a tiny little locking-nut-and-washer, from a pickup truck carburetor. I've been needing this prodigal nut for four months now, and it's more than worth the five dollars I pay Ren for finding it. A good omen for tomorrow.

. . . to build! That is the noblest of all the arts.

LONGFELLOW

STANLEY
No 199

KNIFE

The Penny Knife

Living in a house while remodeling is much like running a race while tying your shoelaces. It can be done, but it takes coordination, skill, and luck to tinker with the rooftree you're living under. But some things are imperative, and this newest job falls under the heading of structurally urgent.

On the south wall of my house, I must place four posts to support a beam (one that should have been installed over a dozen years ago; long story); said beam to carry the weight of the roof above a long window. Simple enough. That means cutting out the siding to inset the posts on the same plane. Down near the bottom, there isn't enough clearance for a power saw, unless I want to tear out the deck, which I don't.

So what I must do may be simple, but it won't be easy: cut off a section of vertical-groove siding, in place, with a knife. This is perhaps easier than using a spoon, but the choice of weapon is limited to one type of knife. My retractable drywall knife is sharp enough, but the thin blade could snap off. I can't find my vinyl knife, the one with the blade curved like an eagle's beak, perfect for cutting carpeting and vinyl flooring (and, as literary buffs know, the favorite skinner of Hannibal Lecter). I must have loaned it to someone. In all likelihood, it wouldn't suit anyway. The carbide knife, a one-tooth saw that is no more than a chip of silicon carbide welded to a handle, excellent for cutting acrylic and laminated plastic such as Formica, would probably do a perfect job. But it makes a thick kerf and would take twice as many strokes. My trusty pocketknife blade doesn't lock, a real hazard to fingers. So I'll have to use my very best marking knife, a bench knife that is terrifyingly sharp and strong as a Bowie.

I place the point of the knife in the groove of the siding and begin scoring from top to bottom, cutting deeper with each stroke. What I really need is a laser cutting–pencil, or stronger arms, or a time machine to go back to 1981 and put in the beam. Remodeling a home while living in it is easy and cheap compared to refitting an aircraft carrier in the middle of a sea battle. The only two impediments are money, time, weather, tools, materials, a rapidly aging body, drop-in visitors, scheduled visitors, other commitments, previous bad workmanship, and my own inertia. I guess that's more than two.

On the thirty-seventh stroke, I cut through the siding. Only seven similar cuts to make. Time to sharpen my pencil for laying out the posts, and thereby give my spaghettified arm time to recover.

In the science of working wood, you must keep your graphite marking wand/word processor sharp. The tripod relationship of knife-to-pencil and pencil-to-carpenter is basic. Measure to the nearest sixteenth of an inch, mark to the eighth, cut straight, and in theory, what could go wrong? However, a dull pencil means a fat mark, which translates in practice to an imprecise cut. Imprecise is wrong.

My pocketknife, an old electrician's Barlow for which I paid one penny, whittles away shavings of cedar and carbon, but it's fairly dull at the moment. Even when the edge is honed, it isn't the best tool for sharpening pencils, but it can strip the insulation from wiring, whittle the tiny curves of a coped joint, cut the backing of insulation, and a thousand other general cuts. In addition, it fits in the front pocket of my denims. My wife found it in a Christmas-tree field, and I gave her a penny for it.

However, I can't always get to it when I'm wearing my carpenter's belt, so at those times I usually employ a drywall knife with its retractable blade. Thumb the blade forward and pare the pencil to a fine point, which incidentally smears graphite shavings on the blade, lubricating it for an easy slide back into the body. This is too symbiotic not to notice.

Pocketknife, rock knife, and marking knife: these are the three main types of knives associated with carpentry. The standard pocketknife can be any two- or three-bladed knife, Swiss Army, Barlow, or Boy Scout. (Electricians have a special fondness for this kind of blade. As a class, electricians have good fingernails, and they can get the blade out quickly.) For most of the last century, a pocketknife was the common knife of carpenters. Long ago, woodworkers and cabinetmakers used something called a Sloyd knife; this is basically a bench knife with a fat handle for fine carving, general whittling, and cutting marks.

But then came the utility knife with replaceable blades, commonly used to cut drywall and hence called a "rock knife" by some. It is the knife of choice for all drywall hangers, most finish carpenters, and even a few carpet layers (especially the younger ones; older rug-stretchers prefer the wickedly curved vinyl knife, which they seem to hone hourly). The utility knife is a recent arrival on the tool scene, the kind of knife most production framers now carry, although timber framers (a different breed of builder altogether) tend to favor a folding belt knife. Buck, Schrade, and Opinel make fine knives of this type; I've owned all three and used them a lot when I built log houses. They were purchased new, and cost considerably more than one cent, although that's what I sold them for.

The marking knife is an oddball, the tool of fine woodworkers by way of the glazier's trade (or plasterer's, or paperhanger's, depending on whom you talk to). Evolving from the original Sloyd bench knife—from the Scandinavian word *slojd,* just like it sounds, a system of manual training for woodcarving—to a more modern (and sometimes very elegant) striking knife, it is the ideal tool for making fine marks on wood, such as the pins of dovetails or the shoulders of tenons. Some called it a "drawing knife" because they drew a V-groove in wood for a handsaw to follow. Mine is also called a mill knife, made by Dexter, with a block aluminum casting and walnut inserts. The blade slides through the body, protruding from both ends, fixed where I want it with a set-screw. Boatbuilders use this kind of blade and leave them out on their workbench where everyone can admire it. Do you know someone who is building a canoe? Look on her bench and see what kind of knife you find.

There's always a pocketknife in my pocket at any given time, and an expensive marking knife or three around my workbench. (Although I paid only one penny for my favorite bench knife. No doubt it was quite expensive in the first place, is what I mean.) But whenever I put on a toolbelt (or any time I am forced to hang drywall), the knife I use most often is the rock knife. Six or so keep each other company in a small drawer in the shop. The triangular blades of this utility knife can be changed in under a minute with a screwdriver, or even a penny. So it's always sharp, and one can hone the dull blades in the evening when the work is done. (I purchase utility knife blades in packages of one hundred, and religiously hone the dull hones all at once, every decade or so. The steel is still good, even if the edge is dull. Only a pig would throw them out—unless they were really trashed.) The utility knife is pure utility, pure function: push the button forward and the blade is right there. In addition, the retractable blade makes for safety. It was the type of knife we gave our daughter, Serenity, for her first tool pouch. I charged her a penny for it. Came right off her allowance.

An eco-alert: One recent version of this type of knife features a scored sectioned blade; you snap off each disposable section as it becomes slightly less than razor sharp. Machines are not the only ones capable of stropping a razor edge on a knife —it's a simple skill—but apparently they're the only ones with enough time to do it. In my considered opinion, the planet needs more tiny fragments of discarded razor-sharp steel like it needs another ozone hole.

Here's an easy way to sharpen a knife or edged tool of any kind or type: First, discard the entire concept of easiness and prepare to spend an hour of your life making a dull edge less dangerous—and a sharp tool really is less dangerous. Go out and buy a three-sided Multi-Stone from Norton, or a grinding wheel made of sandstone that sat in a farmer's yard back when airplanes had two wings. Or get a simple carborundum stone, soak it in oil for a week, and lay it on a block of wood that has been mortised to hold the stone. Or buy every honing stone you ever come across, including the fine-grained stone that barbers used to use to sharpen straight razors. I have done all of the above. The point is, round up at least one good stone. A dull knife is at least as dangerous as a dull chainsaw.

Now move the blade across it, holding it at an angle. You want more detail? It's as easy as chewing food. People have been sharpening knives since the Bronze Age. Let me assure you, there must be five thousand articles out there on sharpening the edges of knives and other tools, all written in the last twenty years. It seems that any time a magazine's readership flags, or any time a publisher wants another title, all that is necessary to catch a reader's eye is to assign some hapless freelancer the job of explaining how to sharpen a knife, with illustrations. If I have read one article on the subject, I have read all five thousand of its brothers and sisters. Each writer says about the same thing.

I say, sharpen your knife however you like. Use a circular, figure-eight, or straight stroking motion. Don't cut yourself.

For the beginning carpenter, until he or she establishes some sense of purpose, a carpenter's knife appears to have one primary use, that is, to keep one's pencil sharp. "A dull pencil makes a wide line, or a fat caret. That's no damn good." This epigram came verbatim from the lips of Swanee, who sold me my first utility knife for a penny.

Thereby hangs a short tale, as well as an explanation for all these penny knives. When I first began swinging a hammer and making marks on boards, I spent a lot of time sharpening my carpenter's pencil. It wasn't necessarily dull, but it gave me something to do that seemed to have a purpose related to carpentry, and therefore something that I wouldn't get yelled at for doing while I tried to absorb the gist of what was going on. Swanee would measure, mark, and cut a board, and then install it in the place where it was supposed to go. Everything was a mystery to me. I watched, carving my pencil. It took a long time to understand all the whys of carpentry; at nineteen, in the study of three-dimensional visualization and spatial relationships, I wasn't exactly a rocket scientist.

One time—this is painful to relate, but it's the truth, so here goes—one time I noticed that his pencil was getting dull, and having nothing better to do, I picked his pocket and began to sharpen it. A minute later he reached for it, fumbling and coming up empty. He started looking around on the ground. I cleared my throat: "Here you go. I sharpened it." He took the pencil, staring at me. Moments passed while an amazing range of expressions crossed his face: disbelief, annoyance, wonder, pity, and finally, resignation. "Thanks," he said.

His helpful apprentice must have been a difficulty to him, but he bore it well. As stated, he sold me my first knife, but it wasn't as mercenary as it sounds. It was pure superstition.

On the first day I worked for him, I didn't have a blade of any kind; he had an extra utility knife, and placed it in my hands. "All right, now give me back a penny," he said. Wondering what the hell, I forked over one cent. "There. That knife's paid for, so you keep it," he said carefully, explaining: "This way, the knife doesn't cut the friendship—not that you and I are friends, you work for me, but that's how the story goes. You should never give or get a knife for free. Always take or give a penny for it." I got the impression he took the whole thing very seriously.

In 1981, a pal of mine was on the verge of going on a world trip, and he was trying to find a suitable folding knife. At the last moment, I gave him mine, forgetting about the penny I was supposed to collect. He took off, and I've never seen or heard from him since.

Only 259 more strokes to go. The marking knife works all right, but oh, my aching arm.

Superstition is the religion of feeble minds.

EDMUND BURKE

HAND SQUARE

A Matter of One Degree

It had been a cold, wet, vile, diseased, and seemingly endless winter, but the first few pale streaks of blue in the firmament made it seem as if the world had not altogether forgotten how to do colors. In that pseudospring many years ago, the ocean storms took a week off, little birds whistled hopefully in the naked branches, and the air took on the peculiar shine that makes the gullible take down their storm windows prematurely. But we who could read a calendar knew that winter would go wretchedly on, sea, sky, birds, and shiny air notwithstanding, for at least another month.

Even so, life was good. I was learning something new every day, working in the comparatively warm and dry (if often filthy) world of house remodeling by the sea. It was a far cry from my recent life as a framer. My hands knew the tools of my trade: hammer and nippers, awl and nailset, plane, square, and pencil, all riding easy in worn leather pouches on my belt.

By faith and good works, I was still among the employed, working for Horace, whose beetling dark brow and minor underbite made him look somewhat like a werewolf in recovery. He ran a full crew through the summer, keeping two carpenters to work with him during the slack season. That winter, one of them was me.

The other guy's name, as I recall, was Gilbert Bear, tall, dark, and laconic almost to the point of muteness. He had a habit of shaking his hammer to signify an emphatic *yes*, as if he were tolling an invisible bell in front of him. For *no,* he tapped on his hat and shook his head. I never found out what planet he was from, although perhaps he told me in sign and I just never understood. Gil kept a tight rein on his tongue, but on the other hand, he could communicate yes and no from as far away as you could see him. This is useful around construction sites.

The fourth member of our crew was Horace's dog, a mind-reading German shepherd named Boris. He was not the ordinary companion animal who rides in a contractor's pickup and sleeps all day until work is over. He was continually underfoot. Whenever

Horace spoke to anyone—often, a split second before Horace said anything—Boris would lock his eyes on the person as well, gazing reflectively until the conversation ended; at that point, Boris would go about his canine business.

Dogs are loyal by nature, but Boris took it to extremes. If Horace kicked a flat tire, Boris would lift his leg on it immediately afterward. When Horace had a cold, his dog looked miserable and rubbed his muzzle. If Horace had something to say, at least one member of his crew would hang on every word. It was a bit unnerving, watching that dog trying so hard to emulate Horace and absorb his wisdom. But so was I, for that matter.

Horace was a genius house doctor. Along any coast, a good maritime contractor knows how to seal out water, the primary killer of houses. But Horace was a specialist, and he waterproofed his wooden patients like submarines. With no wasted motion, he could jack up a house along one side of a foundation and surgically remove a rotted sill down to the last sliver of bad wood, showing us why and where the rot had started and explaining how we could prevent it from recurring. Horace understood gravity and moisture, how they worked together to bring down houses. In foul weather, we worked outside under a plastic canopy of his own design, carefully grafting in pieces of treated wood and new aluminum flashing, weaving repairs into the exterior wall, and finally matching new siding or shingles to the old.

As a science, remodeling can be a living laboratory in which you see how houses are put together, or should have been. When repairing the ravages of insects or weather, you will encounter the ideas of the original builder, noting both errors and good practice, stupid mistakes or neat solutions. Sometimes you get to invent new methods of repair, staking your professional reputation on how long it will last. Horace had done this, many times, and won his steady customers in the process.

Have I said he was an odd fellow? Horace did almost nothing in an ordinary way, and even his tools stood out as fairly weird. For example, he drove nails with a sturdy old Farmer Fudd octagonal-head joiner's hammer, a rare 24-ounce hybrid that is half framing hammer and half engineer's hammer. You see them sometimes in lumber mills or on wheat farms, used for banging on cantankerous machinery. He also carried a try square in his tool pouch.

TOOLS OF THE TRADE

By this time I had formed several well-considered opinions on matters of craft and technique, and I was not shy about sharing them. This was the blush of my fourth year as a working student of carpentry, four years being the traditional apprenticeship required to reach the rank of journeyperson, whose work can be called journeywork, that archaic and beautiful word. It properly means four years under the tutelage and guidance of a master carpenter; I could count only two with my first benefactor, Burdell Swanson. Swanee had given me standards and expectations, and a strong sense of mysteries waiting to be solved over a lifetime of study.

Horace was completing my education by confusing my values and questioning my opinions. "No, a try square is much better than your combination square," Horace insisted one day. Nearby, Boris cocked his head, listening to the argument. "Simpler. No moving parts. See? Can't possibly go out of square, like yours eventually will." Boris let his tongue hang in a smile of agreement: *He's got you there.*

No doubt Boris thought it made sense, but to me this was patent heresy. Swanee had carried a combination square in the pocket of his white overalls, and my first boss had more or less flatly stated that a combination square was the tool real carpenters used. Nothing had been said about try squares, except in the context of manufacturing one's own boards, a mere carpenterial footnote, and checking the ends of lumber from the mill. One could also square up hand-hewn beams with a try square, and create dimensional lumber out of logs.

"Well, you're both wrong," Gil Bear chimed in with his third comment all week, "because mine is the best." Horace and I stopped arguing to heap abuse on Gil's personal square, an aluminum triangle with angles stamped into the metal. We agreed it was functional, although having no class, history, beauty, or worth. (It made an excellent saw guide, in fact, better than Horace's square or mine. Purist that I was then, I could not picture ever owning such a minimalist square, let alone using one. At this moment, there are two in my shop, and I'm going to buy a third. People change.)

Of course, much of my preference for the combination square was nothing more than habit. I liked the feel of it, its balance, versatility, and essential rightness. It slid smoothly into my left hand from its specially made leather holster, doubly protected inside a large pouch. Originally a tool of engineers and architects, the combination square was far too useful not to be adopted by carpenters, framers, cabinetmakers, and just about everybody but holdouts like Horace. A spirit level is incorporated into the body (". . . a little bitty thing, which isn't very accurate," Horace sniffed). A tiny scribe lives in the end of the handle, ". . . hanging on by friction alone, small enough to disappear down your boot and useless for anything practical," said Horace, while Boris just grinned and bobbed his head up and down.

This was all a long time ago. In my twenty-third year I held a rather dogmatic view of tools, imagining them to be either good or bad, useless or perfect. As a matter of taste, I favored complicated-looking old tools with multiple functions and beautiful Art Deco lines. The combination square seemed to embody this ideal. With only two moving parts, it can find both 90- and 45-degree angles. Slap it down on the wood and draw a line, and that line will be square. This tool is fully adjustable with a sliding body (or blade, depending on your point of view) and a knurled nut to lock it. By holding a pencil at the tip of the blade where the slot along one side forms a dimple at the end, you can even use it as a marking gauge.

As you can see, I had my reasons for choosing a combination square over all others. That said, remember this: When selecting a type of tool, people tend to make their choices based on entirely subjective, even irrational, criteria.

In a few months, the currents of my life would take me a thousand miles away, to a drier climate where tools wouldn't rust overnight, far removed from any ocean. Horace and I would never meet again in this lifetime, a plain fact that certainly would have surprised me then, but it stuns me now. *Tempus* does *fugit,* doesn't it?

One day the sun was shining in a particularly convincing fashion, and we found ourselves working on the warm side of a towering old apartment building with advanced dry rot at the grade. The air temperature was an undecided forty degrees, but the dark paint absorbed the sun's heat and bounced it back at us. Soon all of us removed our coats, except for Boris.

When lunchtime came, we sat on stacks of lumber and deployed our coffee, sandwiches, and dog food, the sun on our faces and our backs to that warm wall.

I was relaxing when Boris stopped crunching his kibble and whirled his head in my direction. "Give me your square," Horace said suddenly. He placed it on a 2 x 12 and scribed a line. "Now give me yours," he said to Gilbert. Horace traced that line on the board, and then he whipped out his own try square, clamped it down with his thumb, and made the third line with a flourish. "Now we'll see," he said, reaching for a big and unquestionably accurate framing square. Horace was tired of argument. It was a showdown.

Gil, Boris, and I watched as Horace checked each line. His face fell, and Boris's ears sagged. "God . . . damn," Horace said in a hollow voice. Boris whined and thumped his tail on the deck in sympathy. There was no doubt: My combination square and Gil's triangle were true, dead-on square. But Horace's try square was off by a degree or so.

That noon hour lingers in fond memory: "Gosh, Horace, look at that," I said, Gil Bear sniggering amens at my shoulder. "Why, that's an eighty-nine-degree square. It's . . . it's wrong. Golly. Who knows how long that piece of junk has been lying to you? Guess we'll just have to go back and tear out everything that we've done this winter, until we find the last job when that cheesey thing was still square . . . ."

Horace just took it. For the rest of the day, Gil and I enjoyed ourselves, dredging up all sorts of Horacian quotes on the infallibility of try squares. He could have fired us to cure the noise, but he just smiled through the pain, although Boris wouldn't meet our eyes. Dogs have a low threshold of shame.

However, it was a good lesson in human nature: the next morning, Horace had a brand new try square stuck in his pouch, identical to the first but presumably true. We didn't say a thing. Like us, he just liked certain tools because. That was reason enough for him.

Excellent things are rare.

PLATO

PLANE

The Plane Truth

Joy and I are connoisseurs of old things, which often brings us to places where relics of the good old days are on sale. In the flea market, we find a table of old tools, painted with linseed oil to make them shine, all sales final and forever. I pick up one and utterly lose contact with reality. So out of the blue, my wife wonders, "How many planes do you have, anyway?"

What an odd, arid question. Fewer than a score, probably. Anyway, I've found a block plane that nestles in my palm as if recently born there, its history inscribed on a hand-friendly nickel-plated lever cap that clamps the blade: "Patented 2-18-13." I won't even exist until 1950, a working lifetime later. The price is five miserable dollars.

How shall I tell her of the impatient ache in my hand for this odd little plane? Joy, light of my life—I don't have *this* one.

Five bucks later, through the magic of marital negotiation, it's mine. Scuttling home, I take it apart: cap lever up and presto, the blade slips out. The base of the body, or sole, takes a few strokes over fine sandpaper to shine it up, followed by a light coat of car wax to make it glide. I test it on a piece of scrap wood; the blade has no nicks, but it's dull.

At some point, the tool seems to wake up, sensing an unfamiliar hand. The heft changes somehow. This may all be subjective, of course, or less charitably, a load of mystical crap. But we Celts are suckers for a haunted blade. If I oil, adjust, sharpen, honor, and cherish this plane, someday it might speak to me in the middle of a job. On that day, I'll learn something new about planes.

Pathetic fallacy was the phrase Ruskin coined for figures of speech that attribute human characteristics to inanimate objects, like tools. Being dead, Ruskin is arguably inanimate these days; thus I refute him. It's a fallacy to presume tools don't have

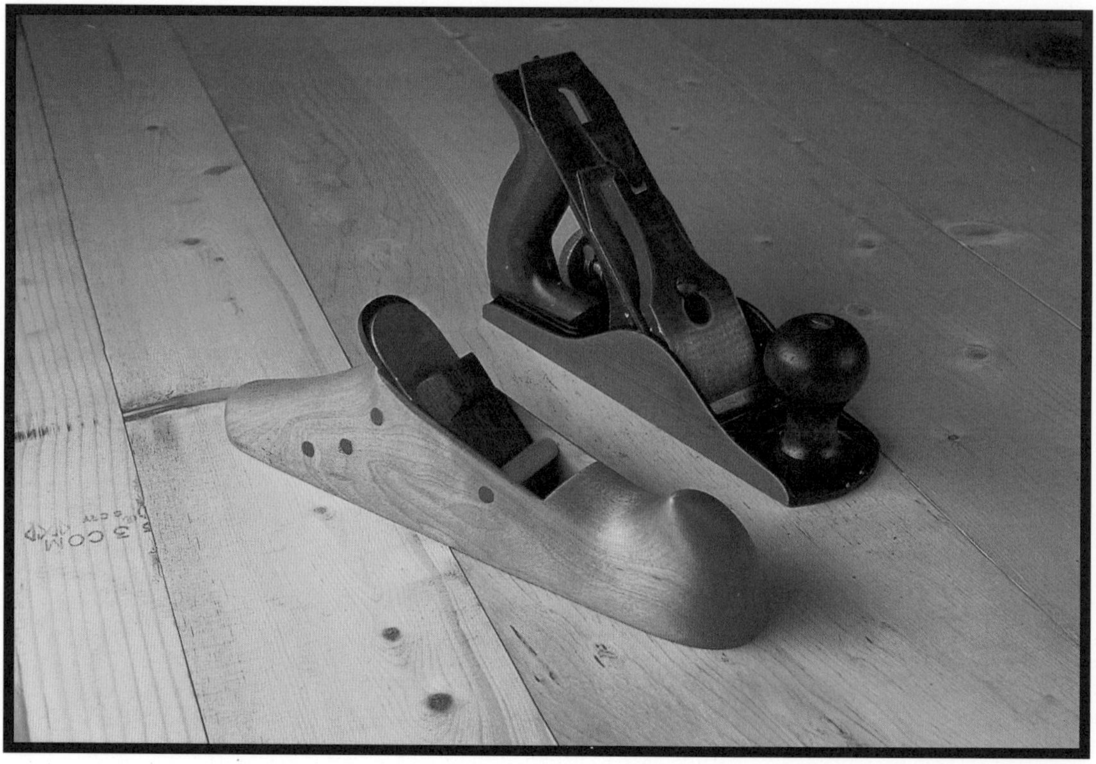

personalities. Some tools are like certain people: You can get along with them. Others, like people, can be obstinate, unfriendly, inert, or electrically hyperkinetic.

Speaking of which. Aside from screaming a lot, my power plane said nothing sensible for years, so I sold it. It was also incurably fast; in my humble opinion, yowling speed is nonconducive to the Zen of planing, which requires silence and patience. If the hand follows the grain obediently, a plane will slice off all the high spots in peace. Skill takes time, practice, and endurance.

After fifteen minutes of whetting on an oilstone and three more on a leather strop, the blade will just about split atoms. A dull edge shows as a fine white line; a keen edge is invisible, slightly sticky when I rub my thumb across it. Those who sharpen handsaws as a hobby are probably taking it too far, as inner voices suggest when I'm fine-tuning ripsaw teeth. However, honing a plane iron to unbelievable levels of sharpness has a purpose beyond pure satisfaction. Dull planes don't work. At all.

After reassembly, I turn it upside down and set the blade a hair (about 1/64th of an inch) beyond the flat of the sole, fiddling with the lateral adjusting lever until the edge is perfectly parallel to the slot on the bottom, called the throat or mouth. Some of the older block planes have variable throats like their larger cousins, the bench planes; the wider the throat, the deeper the shaving. I smooth some scrap pine as a test and put it away—always, always, planes should lie on their sides, to keep the blade sharp. (Photographers often disagree.)

Most homeowners can get by with only a block plane and a bench plane. There are chariot planes, fillister planes, shutter planes, not to mention bullnose, forkstaff, universal, rabbet, compass, coachbuilder's, luthier's, and cooper's planes, literally hundreds of specialty planes; I own a few and can attest. But it's not the common lot to have to work rough wood nowadays, and jobs once performed solely by planes have been usurped by the power router, shaper, and joiner-planer. True confession: I often use a router to make a simple chamfer, a beveled edge on a board, because it's faster and more uniform. After years spent learning planecraft, I can join edges, level a warped table, or smooth and square a rough board. Big medievalist deal, some might say; a novice can do the same jobs with power tools on the first try, after reading a teeny little manual. I submit that reading a pamphlet on human reproduction will not make one an expert on sex, let alone the opposite gender.

Planes have had entire volumes devoted to their mysteries. A Roman invention, this tool has been around ever since, and could survive even the collapse of this electric global civilization. Not that we'll fizzle out like the Romans, who lacked our high technologies, wise environmental policies, and nuclear capabilities. No doubt electricity will always be dependable and cheap, barring a few cataclysmic likelihoods. Still, it might be a good idea to learn how to use a plane.

No other hand tool will torque your patience so quickly. A dull and maladjusted plane is brutally uncooperative; it shaves stomach linings and peace of mind, but not wood. Infrequent users think of juddering a plane across the face of a board, leaving a trail of gouges.

But watch an expert using a well-kept plane. The strokes are slow and even, and the edge of the blade peels up shavings that curl in on themselves as they form. Slowly, a rough or winding board is sculpted into square shapes of buttery smoothness.

The Romans are gone, but planes are still useful for squaring boards, tapering table legs, smoothing rough wood, and fixing a sticky door, probably the same tasks they used it for. On the modern front, there's even a type called a Surform that's perfect for shaping auto body filler. (However, it's not a true plane; more like the ultimate rasp, with a blade that functions like a carrot grater.)

The block plane is the one type that professional carpenters and weekend woodworkers use most often, these days. It fits in a carpenter's pouch and leaves one hand free to hold the work or answer cellular phones. The next most common is the smooth plane, a member of the bench plane family. (Bench planes are so called because a good workbench has a clamp; those who don't clamp the workpiece must devise new scatologies for what inevitably happens.) At 10 inches or so, the smooth plane isn't long enough to take out dips and bumps as well as the jack, jointer, or try planes, but it will dress rough lumber and produce a reasonably flat surface.

Bench planes require two hands to operate. You'd think there would only be one grip possible, but no. One method works like this: For better control when shooting the edge of a board, as freehand planing used to be called, place the thumb of the left hand (reverse for southpaws) on the body of the plane behind the forward knob, with the other fingers on the sides of the plane near the bottom to guide it. Old woodworking manuals teach this grip, but the newer ones advise wrapping all five fingers around the knob for safety's sake. Two equally valid techniques are presented here, so choose carefully.

Unlike a block plane, which has a single blade held down by a lever cap, a bench plane comes with a two-part blade: a plane iron that stiffens the blade and makes the shavings curl, and a cutter iron, the actual edged blade. These are mated face to face with a cap screw, and the whole subassembly sits on the frog, an angled metal rest that comes up from the body.

Here's something worth knowing about plane irons: in time, you will come to hate them. It is tricky to put them together, leaving only 1/16th to 1/32nd of an inch of blade edge protruding. Try it. (Fail. Repeat until successful.)

Swanee, who showed me how to use a plane, was startled by my natural ineptitude as I struggled with my first board. He ripped the tool out of my hands, reset the blade to a fine hair, and turned the board around the other way. I was planing against the grain with the blade set too deep, he explained. "Too much iron. Try it now." I did, and that was the first time a plane ever spoke to me: *Much better, you young dummy.*

A brief distillation of basic, hard-won plane knowledge: When planing a glued-up board, remove the excess glue first with a chisel, because it clogs plane blades. Before the first stroke, locate the high spots—"proud wood," Swanee used to call it—with a straightedge: a square, metal rule, even the plane body itself (by tilting the plane on its side forty-five degrees). Mark any highs with a pencil, and keep checking as you remove material. Planing against the grain makes the surface rough; planing with the grain makes it smooth. When two edges of a board are to be jointed for gluing, they should make a tight joint at both ends because end wood dries faster, and as it shrinks it causes the joints to open at the ends. You can even have a slight hollow in the center; the clamp will close it up, and the glue will hold it.

A final tip: To plane end grain, clamp a waste board behind it so the edge won't chip and ruin your composure. It's a better method than planing from both ends or cutting a chamfer at the off end. Low-angle block planes work best on end grain if you have to trim a lot of end grain, which you usually won't. They're also excellent for hardwoods and miter trimming.

If the wood is cross-grained, you can true up the surface by planing across the grain, but this dulls the blade faster. No problem; sharpen it more often.

Here's a mantra to chant while learning: Planes shave wood. That verb is the best way to think of planes. The plane does not cut, it shaves. You can remove half an inch of stock with a plane, if you're crazy about excelsior or own a hamster. About a quarter inch is the optimal limit to plane off; more than that, consider running it through a table saw or joiner.

Proficiency arrives in three stages: (1) first attempts, (2) sudden comprehension of the difficulties, and (3) competency, after a little practice. Learning to use a plane is as hard as eating one peanut. If you have that much willpower, you'll keep at it, learning as you go until someday a plane whispers to you. Perhaps it will quote Ruskin: *When love and skill work together, expect a masterpiece.*

If it is not true it is very well invented.

BRUNO

HANDYMAN

The Bequest is a Question

The family legend goes like this: all male Taylors, many generations back, were exacting carpenters and builders. Taylor is the twelfth most common name, so it follows that one branch, one tribe may have been so gifted. Indeed, Augustus Deodat Taylor was an inventor of the timber framing method, and he may be a distant ancestor. Or not.

But I was first exposed to carpentry by a relative whose youth was spent stacking sandbags on the isle of Saipan at a furious rate, prior to the largest banzai charge in history. He wanted his jobs to be done before darkness fell, no matter what, and he worked as fast as if his life depended on it. Even as an infant, I sensed something wrong with his methods, tools, fasteners, building philosophy, materials, schedule, and temperament as he worked on the houses we lived in. The world's worst carpenter: Let him remain anonymous, for he is my father.

These days, to the horror of his doctors, Mac still builds at top speed, and accepts as fact that all jobs other than the construction of Rome can be completed by nightfall. Only now he has a reason to hurry.

We still work together occasionally, on his house or mine. It's galling to see him hold a piece of wood in place, make a nearly invisible mark with a fingernail, cut it freehand using a dull saw, and somehow obtain a perfect fit. It can only be explained by genetics or phenomenal luck. I've seen him jack up a house with levers and blocks, dig a footing with a broken kitchen ladle, lay in two rows of concrete blocks with a tiny pocket level, and thereby create a retrofitted foundation that is not off by more than the thickness of a pencil. His tools are a motley collection of rusted junk. Whenever he locates a decent one, especially a precious antique that the gods would never throw my way directly, he gives it to me. Or I steal it from him and leave a new one in its place.

"Did you ever find out," my father asks me now, "what that tool is called? The thingy with all the bits?"

This is an old game with us. He always asks about it. Let me sketch in a little background: We are standing outside his house in the drizzling rain, his favorite working weather. We're putting up sheets of siding, which must be installed under these punishing circumstances for reasons I'll come to. My wife and mother are inside by the fire, like intelligent people. Growing up, I used to hate working like this, always under conditions of combat-grade field expediency. But I'm past forty, and now I see the fun of it. Also, it could be the last time we work together.

"Oh, sure," I grunt, pushing up the sheet while he eyeballs it for level. "I looked it up. It's a blimpwright's gondola bruzz. Extremely rare."

"You lie," he says happily, enjoying the filthy weather. He thinks a moment. "It's a bodger's travice-and-sprig auger." The man is a crossword-puzzle wizard, and who knows for how much longer? For the last two years, he's had a six-letter word for terminal. "Wait. Gotta be: A clogmaker's sabot screwdriver."

A drop of rain finds its way deep inside my ear. My turn: "Okay. I've been keeping the truth from you all these years, just so I'd know something you didn't. It's actually a ship-chandler's baleen fid. Came off the *Ahab's Leg* out of Boston." I hold the siding in place, with my knee, until he tacks it at the top, and then we trade places so I can nail it off—over the studs, evenly spaced, correctly; not his way.

He's a lot more careful of his work than he used to be. It's nice to see your father's skills improve. It makes you proud that you had some effect on his character during your formative years.

It's passing strange that my life took this direction. One of the awkward things about having spent the last twenty years as a carpenter is how completely I failed to predict an accurate career destiny. I envisioned the future with myself in the picture not as a writer but as a latter-day house builder, flying a hovercraft pickup truck loaded with laser saws and anti-gravity scaffolds, a computerized contractor. But my unconscious idealized image of a carpenter always stood beside a buckboard wagon, looking wise and sharpening something in a bare field around 1890.

There were traditions to be carried on, if I could ever find them.

And then I met Swanee. It felt like an omen that our first meeting occured beneath the rooftree of a Victorian-era Carpenter Gothic house under reconstruction. Perhaps he sensed in me a certain rudderless denial of contemporary values.

Good old Swanee. My soul bears the mark of his stamp, and my ears still ring with his instruction. His method was Socratic irony, sometimes with every syllable dripping sarcasm. He liked to ask me questions about what I was doing, listen with arms folded to the answer, and then demonstrate its naivete. At one point in our association, he stated his reluctance to tamper with my nearly seamless ignorance. Only my obvious passion for tools persuaded him to do so.

From next door come the gentle strains of Steppenwolf. I slip into a timeless state of grace; "Magic Carpet Ride" is one of my favorite songs for life. But not my dad's. "What's that awful noise?" Mac looks around until he locates the source of the thumping. "Why are you smiling? Are you on drugs?"

We've been arguing sixties music and politics for thirty years now. "This is good stuff. You want 'Tuxedo Junction,' move to a better neighborhood. Tommy Dorsey still sucks brass," I tell him, a dab of heresy to get the conversation going and maybe warm up the day.

"Your brain has been fried by amplifiers," he says. "Too late for you to lose your tin ear." Next up is "Bad Moon Rising"; he rolls his eyes, but starts to rock out before he catches himself. It's an unfortunate choice, given the lyrics: "I know the end is coming soon, hope you are quite prepared" and so on. The radio DJ switches to "Uncle John's Band" by the Grateful Dead. He never liked them until he found out the name of the band. "Ah, George."

"Jerry."

"Whatever. That's not too bad. I can listen to that."

"There's hope for you yet, Mac."

By the time Swanee died, I had acquired a somewhat schizophrenic approach to carpentry. My power tools were the best and most expensive I could afford, while my hand tools tended to be antiques made of rosewood and brass, found in museum condition for a few bucks at sales and junk stores.

Years later, disillusioned with hard work for low pay, I took employment as a Corporate Tie Person, a cologned and shaven Beamer-propelled technical writer in the homebuilding industry. For a few years, the clean, warm, and well-paying indoor fields of architecture and engineering suited nicely. But some inner warning told me to throw down my phone and pick up a hammer once more. I went back to work. Blisters had begun to form again on my soft hands when the entire conglomerate and all its holdings went into receivership, leaving thousands jobless and wailing. A close call, was that secure indoor job. My fate as a lifelong all-weather carpenter was sealed.

I will probably never stop building on houses, nor ever rise to the rank of Executive Tie Person. Blame my father, all the unanswered questions of his life, and the nameless tool with all the groovy bits in the handle.

"We should probably hurry this up," my dad remarks, heedless of the fact that we're already soaked. Pneumonia doesn't scare him one bit. He just wants to get done before nightfall.

"Bullshit," I say. "I'm gonna wait for the Mighty Quinn." This is code, but he knows what it means. Taken from a lyric in the proto-reggae song of the same name, it means I intend to embrace the Jamaican work ethic, and speed be damned.

What's that tool called, anyway? Its hollow handle is made of rosewood. The knob at the end unscrews three full revolutions on wooden threads, and ten different bits spill out: a variety of slotted screwdrivers, a tack lifter, two kinds of awl, a gouge, chisel, gimlet, triangular file, and even a short saw blade. These steel square-tanged bits fit snugly in the checkered chuck, and the tool rides everywhere in a thick leather pouch on my toolbelt.

I love it for many reasons: the way it feels in the hand, the variety of jobs it can do, and its timeless beauty. In terms of pure utility, it's not very efficient when compared to modern cordless screwdrivers. But my father gave it to me long ago, more as a talisman than a tool, and you can't argue with sentiment.

Whatever it is, it looks at least fifty years old. I own several others that are similar, the oldest bearing a patent dated 1884, but none have the slightest trace of any manufacturer's name. What is it called? A carpenter's interchangeable gouge/awl, a woodworker's multibit screwdriver? It appears nowhere in my extensive library of tool dictionaries and woodworking manuals. I have shown it to carpenters and builders for the last fifteen years and received a multiplicity of different names and uses.

Recently I met a piano tuner who used one, which was exciting until he explained that it wasn't a true piano tuner's tool. Like me, he didn't know what to call it.

It really doesn't matter. There are too many names, as the *Tao Te Ching* tells us. Naming is mere high-level abstraction-cognition. Names are not the things named.

Swanee kept one in his toolchest and didn't seem to think it needed a name. When pressed, he said it might be called a "handyman." But a few years ago, I worked with an impeccable finish carpenter who, like me, carried one of these nameless tools, and she impressed upon me how frosty the walls of Hell would become before she used such sexist nomenclature. We agreed to call it, pro tem, a "handyperson," and swore that the first one of us to establish its true name would call the other. I'm still waiting.

Mac snaps his fingers. "Got it. A hoopmaker's brog."

"No way," I tell him. "Maybe a saddlemaker's punch."

There is something comforting about things unknown. Not knowing the name of a tool means that tomorrow will always hold the promise of learning something new. It may be merely my personal prejudice against the philosophy of determinism, but an unknown future is preferable to the fixed past. Perhaps the only proof against optimism is certainty.

Eventually the rain drives us inside, still batting ridiculous names back and forth. We might finish the job, another day. The doctors claim that he's defied medical precedent so far. Touch rosewood.

FRAMING SQUARE

Carpenter's Calculator

It was almost performance art. A hundred eyes were watching, and they must have felt it. Once more, innocent wood was pulled from the stack. I watched them select three more 20-foot 2 x 12s. A couple of young framers had volunteered to build stairs for the dozen or so buildings under construction in the apartment complex; twice they had bollixed the job, and the other workers had given them a nickname. For the third and final time, the Notch Choppers grimly slapped their framing squares on each board. Once notched, these boards would be called the stringers, and the entire tread-supporting framework the stair carriage. In theory, that is.

The scrap pile looked like a boneyard for dragon spines, but now their radio was turned off and their concentration was frightening to behold. In silence, they measured, marked, and cut. They set the carriages in place. No good; light showed under the bottom cut, and the top cut didn't meet the landing. Sure enough, the cuts were wrong again. Proper terminology for the stringers and carriage was now trash.

Unfortunately, there were witnesses on the balconies, and ironic applause erupted from the electricians, drywallers, roofers, and finish carpenters who were happily collecting on bets. One of the Choppers screamed a heartfelt pair of words at the gallery and cast his aluminum framing square like a boomerang in the general direction of smiling faces. It missed, landing in the dirt a few yards away. In lieu of a formal resignation, he got in his pickup and drove off. His partner found a place to hide until those in charge could locate and chastise him.

The same afternoon, still picking pieces of that guy out of his teeth, the superintendent approached my old boss, Burdell Swanson. Would we consider trying our luck with the stairs?

"Sure, if the price is right," Swanee told him. "The kid and I can probably handle it." The kid was me, his lowly apprentice. We had been hanging doors and other finish work, small tasks that I had begun to master. The framing crew did not hold finish

carpenters in high regard, viewing my employer as a wasted old man and Yours Truly as his dumb helper. They were only half right.

The next morning, I learned what a framing square could do: virtually anything, according to Swanee. His feeling for that tool obviously bordered on reverence, and my ignorant hands were not allowed to profane his black steel square. It was not just a tool for establishing a straight edge or a right angle. It was also a calculator, a slide rule with no moving parts. He tried to explain, speaking in fairly psychedelic language about the rafter tables (Southington's versus Sargent's and the advantages of each), some prolonged gibberish about the Essex Board Measure, and there was a bit on using octagon scales to find simple angel/pinhead ratios and the location of the Titanic. I nodded like a trained horse.

Soon we arrived at Basic Errors, my particular field of expertise. Swanee showed how the Notch Choppers had miscalculated the staircase's total horizontal distance, or run—one of the magic words for operating a framing square. The other was rise, the total vertical distance. We laid out the first carriage, made the cuts, and set it in place. (Swanee did, that is; I observed and helped move them.) The result was, to use Swanee's idiom, "a honeymoon fit." When the treads were installed, each riser was exactly the same height; to prevent tripping, staircase tolerances can vary by no more than an eighth of an inch. Swanee's cuts were probably closer than that.

Afterward, he warned that his framing square was probably the most dangerous tool in carpentry. For one thing, a tiny mathematical error could ruin an expensive board. For another, he said, if I ever accidentally dropped his personal steel square, it might bend out of true, and then it would be his sad duty to kill me.

So my first framing square was the discarded aluminum missile of the departed Notch Chopper. Its flight had bent it a mere one degree off from a perfect right angle, which Swanee fixed by rapping a wooden hammer handle on the soft metal. "Good enough for you to learn on," he said.

It was, and I did. I learned that the long arm was called the blade, and the shorter arm the tongue. I learned about gauges, those little brass button clamps that marked certain numbers on the blade and tongue, and how to use wing dividers and a bevel square as ancillary tools with the framing square. I studied the rafter tables and the alternate (and to me, easier) method of moving the square along a board, called stepping-off. And still my employer insisted that I had only scratched the surface of the mysteries locked inside it. So I kept learning, and eventually discovered a very important fact of life: a tool can be both indispensable and obsolete.

Today I own a number of very old books written by scholars of the square. If you ever have about twenty-five years of idle time, study the framing square. You will either wind up very knowledgeable on the tool or a raving monomaniac. In 1909, one author pondered: "For over 25 years I have hoped that some genius would show the world how handrailing may be laid out by the use of the steel square. I myself have sought the formula through many sleepless nights. I predict that handrails will be laid out altogether by the steel square before the end of the first quarter of the twentieth century, and a fortune awaits the man who makes the discovery."

This is a *cri de carré*, and it would not be polite to laugh. The poor bugger has spent a lifetime besotting himself with geometry and thinks future generations will carry on in a gleaming world of laser buggy whips and steam computers. But the future of carpentry turned out to be assembly of mostly precut millwork, including stairs and roof trusses.

Even in houses where rafters must be cut on the site, many carpenters would not consider using a steel square to find a rafter length. They can look it up in a book of tables. The same applies to stair carriages. In the early eighties, I finally gave up and started using tables like every other carpenter because tables were faster and "to hell with slow" was the motto of the decade.

My fondness for the steel square is nine parts sentiment and one part practicality. I can still lay out rafters, common, jack, or octagon, with a steel square. But this tool can do more mathematical calculations than most people have time to learn. Given the diameter of a cogwheel and the pitch of its cogs, you can easily find the number of cogs with a steel square, or determine the length of a hoop around a wooden tank. Got any pulleys you would like to replace to make a shaft go faster or slower? A steel square is the answer to most calculating problems you'd encounter on a small farm in 1909.

New uses for the framing square are unlikely to bring wealth or accolades from a grateful public. Even so, most carpenters and woodworkers keep one around because they are ideal for adjusting a radial saw when it slips out of adjustment, or for cutting very wide boards. They also make a fine saw guide; keep the shoe of the power saw against the tongue of the square and the blade against the board, and the cut will be perfectly straight.

The framing square is generally made of steel, but not always. Aluminum versions are lighter by several pounds, although some purists claim a strong breeze will put them out of square. Others say that is a rotten lie, so my opinion must remain secret. A few very old and rare models were made of brass, which supposedly does not get too hot to pick up under a blazing sun. If a collector finds one, it will never see full daylight again.

My own square is a handsome antique, a blued-steel Craftsman R–100–B with engraved numbers and Sargent tables. I treat it with the deference due any unsolved puzzle, and from time to time it shows me why the basic framing square, found in Egyptian tombs, will always be with us.

Recently, I used it to make rafters for a small barn, using peeled poles instead of dimensional lumber. I set the poles up on sawhorses as usual and stepped off each unit of rise and run; the joins at ridge and wall were reasonably conjugal. And who knows, my manual of rafter tables might have worked as well, had the mice not found it in my shop and chewed it to confetti.

A little learning is a dangerous thing.

SENECA

HAND DRILL

The Original Cordless

"Like, it comes with a spare energy pack and a charger," the young clerk says, watching me perform a few imaginary tasks. For different reasons, both of us want me to buy this new 12-volt, pistol-gripped, keyless-chuck, variable-speed, high-torque cordless electric drill. A serious machine. The salesdude has done his part; he pauses expectantly, waiting for my willpower to run down. If I whip out my plastic now, we'll both be amped to the max (excited and happy).

However, I have a Q: "Solar charger, or conventional?" He blinks. An eco-freak? Patience, he tells himself. Solar charging for cordless tools is a few years away, he advises. "They're working on it. Soon, no doubt." Like he cares.

The moment passes, and I put the tool back on its display rack. "Maybe another time. Anyway, I already have a cordless drill, sort of. You know, a little 'hurdy-gurdy'?" I make a cranking motion, and he forces a dyspeptic Die-Boomer-Scum smile. The thought balloon over his head reads *The customer is always right. But the public is an ass.* I thank him for all his help and promise to return someday.

Hardware stores are dangerous places full of marvelous tools, without which life can seem much less appealing. Long ago I resolved never to pay more than a hundred dollars for any single tool; only great desire can bend my iron rule, as occasionally happens in the face of too great a temptation. These days, a cordless electric drill is fairly resistible, not least because I'd have to explain the purchase to my wife, Joy, who has a gift for concealing her enthusiasm over new tools. Besides, for that much, I could buy twenty hurdy-gurdy drills.

As a matter of fact, not long ago I ran across a man putting out a sign for a small farm sale. His face was a deeply etched map of laugh lines, and he had a way of running his palm over the top of his smooth head, as if he was brushing sawdust off it. Apparently he had taken leave of his senses, and now he wished to divest himself of all his hand tools. He showed me a few oldies of such polish and patina that little spasms galvanized my solar plexus and arms as I jerked my wallet out. I bought as many as I could afford, with every penny I had right down to the change in my pockets, spending feverishly before some scabrous antique dealer came along and cleaned him out; exactly what happened the next day.

We talked as I loaded up my car. This weather-beaten old fellow told harrowing stories of his childhood in regions so inimical to human life that the legs of his bed stood in cans filled with water to prevent insects from eating him. He also recounted the long working years spent collecting hand tools, the very tools he was now selling off, and he gave me his personal secret for happiness: "Always deny yourself one thing in life. Keep turning it down when it presents itself; pick anything, but stick to it. That way, you'll always have enough." Food for thought, I thought.

Included in my wooden box of purchases was a goodly assortment of hand-operated drills, lovely old things: elderly bit braces, handworn gimlets, wood augers of rare device, Yankee screwdrivers in new condition with complete sets of bits and drivers, breast drills, hand drills, and nickel-plated push drills. Joy wasn't exactly elated, but she admitted I'd gotten good value for the money. Originally the cash had been earmarked for a few nonessential purchases and odd bills in town, to which there was not much point in going since, after attending the mad farmer's divestiture, no money remained . . . .

She-who-must-be-mollified is a goddess of patience who does not often criticize, but on this occasion she seemed prepared to make an exception. "Wait," Joy said. "Now, you spent fifty dollars on what, exactly?"

Note how it sounds like a request for information, but is not. At this critical juncture in such conversations, it is best to make a bold and honest answer. With some pride, I began to tell Joy about the history of our new investments.

For sixty years, from the 1870s to the beginnings of home electrification in the 1930s, the hand-operated drill was the standard tool used to make holes in wood. And then the electrical outlet was invented and power drills evolved. Thousands of hand drills like these still lurk in the bottom of thousands of tool boxes, or lie mouldering on the shelves of forgotten garages. They still work perfectly and will last forever. After all, almost no one uses them in the modern age.

Hand Drill

Over the years, my taste in tools has tended toward the basic, antique, and many, rather than a complicated and modern few; this reflects the dichotomy of living simply in the nineties, when More and Better is barely Enough. (Examine your own philosophy: Are you happy with your current things, or do you anticipate requiring More of them, and Better ones, in the future? Given what you have now, would you like a little Less?)

This tool is definitely Less. It is not what woodworkers call an aggressive tool, requiring a rock-steady hand lest it bore too deeply or quickly into the wood fibers with sheer power. This unpretentious little drill uses simple bevel gears, like an eggbeater. No river is dammed, no atoms smashed to make it turn, and it takes a little longer to do the job, but what's time to a drill? Speaking for myself, the Inner Luddite prefers the small whirring sound it makes as its spiral bit eases into the workpiece, an improvement on silence.

It isn't just unadulterated laziness, but whenever I need to make, for example, a pilot hole for a screw in hardwood but don't feel like stringing out an extension cord, this is the drill I grab. Granted, it's no more than the nonelectric version of the common cordless screwdriver. However, it's a lot cheaper, and it brings a precision to the job of making small holes that is absent with most other drills, electric or not.

One version that uses the same basic principle is the breast drill. Quite a few have passed through my hands, and I finally pared down my collection to one that works like a damn. Largest of the portable hand drills, a breast drill is a quiet, efficient machine you can lean into while cranking, with your body weight putting just enough pressure on the bit to give it a solid bite. One hand cranks while the other hand steadies, the butt-plate resting slightly under the sternum. It will put neat holes in metal, wood, plastic, even concrete, just as well (if not as fast) as an electric drill; an excellent tool, nobly named after the place in humankind where hope springs eternal. With the flip of a lever, the driving bevel-gear can be set in two positions, on an inner or outer ring of gears for a fast or slow speed.

Primitive versions from the last century had an uncomfortable round plate, doubtless very hard on the chest, and only friction or a set-screw to hold the bits. Mine was probably made after 1910, the acme of breast-drill evolution, just before it spiraled downhill to obsolescence. It only lacks a spirit level incorporated into the body, which some of the better models of breast drill featured. I want one. Imagine being able to make a level hole just by looking down at the little bubble, I told my wife. "Imagine," Joy echoed.

Both hand and breast drill have a Barber-type chuck: the tang of the bit is clamped in three jaws, which automatically center the bit as the chuck is tightened down.

One neighbor, a silkscreen artist by trade, told me that his small hand drill works quite well for assembling frames. But "I call her a 'hurty-hurty'," he told me once. "When she jams, my fingers get caught in the gears." He showed me an excellent blood blister, and I told him how to avoid jamming: bore the bit all the way into a block of paraffin before drilling wood. This method also works with all other drills, even a cordful electric, if you're drilling extremely hard wood.

In the interest of full disclosure, a confession: There are moments in the wee dark hours when I contemplate how perfect it would be to sell all my old hand-cranked tools, to part with the vanished past and invest in the invisible future, handily symbolized by the purchase of a beautiful new rechargeable electric drill, fully loaded with keyless chuck and spare energy pack (battery).

Then I think about that crazy old farmer's advice, and fend off the desire a little longer. Certainly it would be a big improvement over my little hurdy-gurdy. Someday I'll pick something else to deny myself and buy that pretty new cordless drill. But I'm almost positive that I'd never again use a hand drill for all the jobs we do together now.

Progress — the stride of God!

VICTOR HUGO

DRYWALL TROWEL

Master of Plaster

Serenity is ten, the age when parents hope their young will discover the concept of gratitude. (Hope is cheap, fellow parents, but thanks are always dear.) We, her loving father and mother, have spent the last week remodeling her room, every day and yea unto the nightfall. We will paint the walls and lay carpet in a few days, just as soon as we are done plastering.

We began by moving our daughter's furniture out of her bedroom (and into the living room for a week, O ye parents). Then we ripped out the ancient carpet, old and funky with fingerpaint blotches and a cat memo from a stray tom she had adopted. But the walls were the worst. Some idiot had mixed corn meal into flat paint for a faux texturing job, creating a surface only slightly less abrasive than sandpaper.

Tonight, we are slowly burying this gritty surface under a better one. Memorize the following: Do not ever become proficient in a skill you will dislike. Learn to do an ugly, unpleasant job of work, acquire a little hands-on experience, get good at it, and you will do it often.

Somewhere over the course of my life, I learned to plaster. I can, therefore, I am. The sun went down hours ago. Our daughter is asleep, and even the television has signed off. Only Joy and I are awake, smoothing ickiness on our daughter's bedroom walls and ceiling but trying to keep it off the floors, at the ungodly hour of three in the a.m.

All my energy for the work evaporated hours ago. "Is there any more coffee?" I ask my wife, who has white goo caked on her face, hair, clothing, and hands. Joy could pass for a survivor of a bakery explosion.

"You've already had three cups," she reminds me, in the wifely tones that mean *Dumb idea, honey.* The first cup was fortified with brandy, which seemed like a good idea at the time. Irish coffee contains the four essential food groups: sugar, caffeine,

alcohol, and fat. It liberates the creative impulse and numbs the busy intellect. However, now I would like something to slap me awake for a few more hours. Sleep beckons, and I must refuse.

True plastering, in the original sense, is almost a lost art, requiring a scratch coat (sometimes called a brown coat) of gypsum plaster over metal or wooden lath, a second coat over that, and a lime-based hard finish coat. It requires a practiced hand. For one terrible month many years ago, I chopped and nailed lath and mixed up plaster for a gentle old Irishman who remembered how to create a wall the old-fashioned way. Mr. O'Hara had the patience of a saint, and in that month he taught me the basics of an obsolete craft. It was hard work.

Later, I took a job as a drywaller for Burt, the recovering murderer who bore scars and tattoos as thick as the graffiti on a New York subway. He had the patience of an itchy rhino, but we got along. Hanging and plastering drywall was soulless work; it was also a lot easier than the old methods.

Mercifully, drywall construction has replaced the old lath-and-plaster. But we can still use the word *plastering* to mean applying a perfectly uniform and pond-smooth finish to the walls and ceiling of a drywalled room. If you don't know how, give thanks to the universe.

Putting up sheets of drywall is the easy part. You hang the plasterboard, staggering the joints and cutting out little squares for switches and outlets, or little circles for the overhead lights. The result is fairly pleasing to the eye, but it is not seamless yet. Now you must tape the joints and fill all the nail dimples. Then you must sand. Apply another coat of joint compound, and you sand again. Go over every joint and dimple a third time with topping compound, and, by Saint Jude, sand everything once more. Now you are ready to texture, or to die.

When I hang drywall, I always tape the joints with sticky fiberglass mesh; using paper tape is a real pain in both glutes unless you rent a "banjo," and you would know why this commercial tool is so called if you saw one. Even if perfectly applied, paper tape can still separate from its bed coat. Two-and-a-half-inch wide mesh tape with a pressure-sensitive adhesive backing works better, in my expert opinion. It requires no bed coat, it goes on in one easy step, and it represents an improvement over paper tape.

Like every other building enterprise, plastering requires the proper tools. The tool in my right hand is a drywall knife, or trowel. In the left hand is a plasterer's hawk, weighing by now seven subjective tons. These two tools will reduce both brawny arms to quivering pipestems after an hour. Plastering is the least fun of human endeavours, at least in the beginning, middle, and just before the end. But it is like hitting yourself in the head with a hammer: it feels so good when you stop.

These and other circular thoughts have taken up permanent residence in my head, a philosophical trap for a sleepy mind. If our daughter were not asleep in the living room, I would put on a Steppenwolf tape. "Magic Carpet Ride" at full volume has gotten me through any number of all-night projects.

The taped joints get two coats of joint compound and a final shot of topping compound; all the burrs and rough edges are sanded out between coats. It is an awful job. One asks—Why? What madness causes humans to sand an entire room, to snort fine white dust six inches up both nasal passages? I'll tell you why: if little granules of dried compound remain on the walls between coats, they will break loose under the trowel and plow grooves in the next coat of plaster.

The most massive wave of fatigue will come at the halfway mark in time, which is also the seven-eighths mark of energy reserves. You can't quit now, or even think about a long break from the work; if you do, you'll abandon the project for a day, and probably another one. This can stretch out into years of procrastination.

Behold the drywall knife. I usually call it a trowel, although it was called a plastering knife in hardware stores for many years. Some people still call it a knife, although anyone can tell that it is not really a knife at all. I own several hunting knives and pocketknives, and have become quite proficient at recognizing a knife on sight. True, the drywall knife has a blade, but so do an axe, chisel, and framing square, three more tools that plainly are not knives. So let us call it a trowel, instead.

This tool requires a bit of breaking-in. Using a grinder, remove the sharp corners of the trowel, rounding them off a little, no more than an eighth of an inch. Then bend the blade ever so slightly, so that one face of the blade is convex and the other is concave. The convex side usually goes against the wall when I plaster, so that the tips don't leave ridges. For taping joints, I flip it over, using the concave side to apply joint compound. The compound shrinks as it dries; leaving a minor, nearly invisible bulge at the joint helps to produce a flat surface.

This is the proper technique: Scraping a bite of plaster off the hawk, you run it along the joints in one unbroken motion, keeping the handle at a 25-degree angle to the wall. There should be a little excess plaster left at the end of every stroke; wipe it off on the edge of the hawk to clean the trowel. For the first coat, use an 8- or 10-inch trowel. Switch to a 12-inch trowel for the final coats. However, before you can prime and paint the walls, you must first texture them.

A friend of mine once rented a machine and proceeded to discover that he knew nothing about blown-on texture. In desperation, he found a comb and began brushing it into patterns resembling a bad mistake. But he kept working, trying for that random uniformity that marks a hired professional's work, and by the time he was done, it looked like ratty sheepskin.

There is an easier way to plaster and texture, requiring only two coats, and with practice, the result can be quite professional-looking. It seems to work best on old walls with bad texturing jobs. But be advised: It is an experiment, a method I've used for only ten years without any failures. Maybe it will flake off within another ten, so I can't recommend it yet. But I will tell you what it is, and if you want to try it, lotsa luck.

Glop two and one-half gallons of premix joint compound in a five-gallon bucket. Dump in ten cups of silica sand. (Warning: Wear a mask and take care not to breathe the dust.) Add one pint of water and one cup of latex bonding adhesive. Using a large cage-type or bladed paddle on the end of a large drill, mix everything to an even consistency for one minute. I use a monster ½-inch drill, because it has enough torque to turn the paddle for two minutes without burning out, as did every ⅜-inch drill I have ever used.

Still with me? Now, if you wish, you can plaster this directly over the first (dry) coat of joint compound, which you have already primed, right? Using a wide, broken-in drywall trowel, float it all very smooth (practice on a closet wall). Let it dry, but do not sand afterward, or ever; just scrape off any ridges and then prime and paint. The result feels like a plaster wall, but without the expense or deep unpleasantness of lath construction.

Please understand that, for perfect results, you should probably apply this over a special drywall board with a porous paper facing that adheres to plaster without separating from the gypsum core. Some drywall manufacturers offer complete veneer plaster systems that hybridize drywall boards and plaster, and for new construction, that's the way to go. But my method seems to work well over old surfaces, and new ones if they are primed first. Time will tell.

A bright light appears in the eastern window. Dawn: Time for a bath. And then bed, delicious bed. We'll be finished with Serenity's room this week. She'll probably thank us, once or twice. That's all right. The real gratitude will come when Serenity has children of her own.

Wash you, make you clean.

ISAIAH I.16

HANDSAW

Why, Daddy?

In this season, even electricity seems to flow like frozen treacle. As I am busy writing this sentence in my cozy sanctum, somewhere a transformer goes *Ssnzrt-POP!*, a victim of target practice, cold weather, bad driving, or a squirrel deleting himself from the evolutionary chain; whatever cause it may be this time, the net result is that electrons stop flowing into our house. My computer reacts: *Thud, Zip! (scrunch)*—C:\IRRECOVERABLE DATA LOSS, PAL. And all the words I was going to tell you about handsaws have now disappeared into cyberspace. To commemorate the event, I speak an antique Word.

In the other room, Serenity leaps off the couch with a happy cry, flipping open a small notebook she's been carrying. "Ten more ce-ents," she trills, writing it down. "Guess what, Dad? You're almost up to a dollar."

Really? It's been one of those mornings. "Sorry. Put it on my tab." Oh, s—t. I really must hang a soundproof door on my study someday.

Nevertheless, this new method works, enhancing her allowance by helping me monitor my common language. For many years, I worked in a trade where unwished events of any kind usually elicited the kind of earthy mots you just don't want your kids muttering when their Nintendo characters die. But fair is fair; the fault is mine, and now I must pay for my unedited speech, a dime at a time. To give her credit for initiative, Ren thought of it. The chart she drew up, listing bad words and the corresponding fines, is a miracle of phonetically spelled profanity. I will treasure it forever.

We misnamed her Serenity, on the morning she was born in this very house. Someone once said that having a child is like having a bowling alley installed in your head. This is certainly true of small children and later, I am told, of teenagers. But when they reach age ten, they attain that quietly hypercritical plateau when they seem to examine you at odd moments, quoting back fragments of your own spoken philoso-

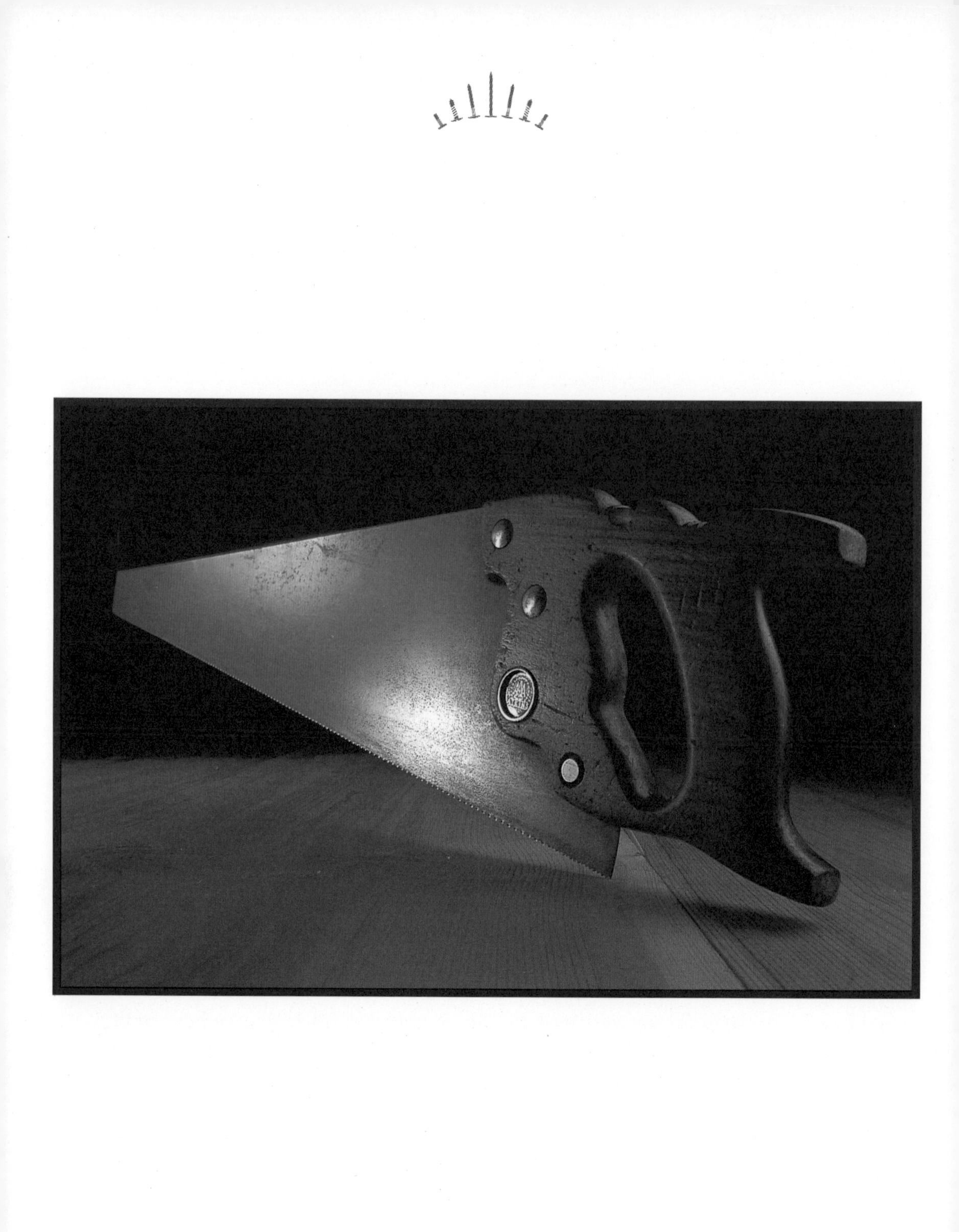

phy at awkward times. It is much like having a tape recorder running constantly, sometimes while you are merely thinking aloud. Consequently, you learn to think aloud less.

Time to take stock of the morning. With the power off, I can't be expected to write today. I might jot down a few ideas in my little electronic notebook, but aha, the internal battery seems to be dead. Another word gets as far as my glottis before I choke it off to a whisper, for economy. The universe seems to be communicating, advising rest. I'll bet I'm supposed to loaf and think and read and cogitate all day.

"I'm bored," Ren says, standing in my study doorway.

"That's absolutely impossible," I explain. "Read something. Start with *The Complete Works of William Shakespeare.* That ought to kill an afternoon."

Kids are so old these days. Ren gives me an untellable look, sophisticated beyond her years. "Dad—I'm bored because . . ."—she lists off all the things she can't do for one reason or another—". . . and it's raining, Jenny's at her grandmother's, Mom's in town all day, the TV won't work. . . ."

The list keeps growing until I flag her down. "All right. Enough. Let's build something together. That's why we have a shop."

Fortunately, the shop has skylights and a woodstove, free heat and light, independent of electricity cables. Ren and I get the fire crackling with wood scraps, and we decide to create a birdhouse. This is a simple project that any father can make with his daughter, a miniature domicile for feathered bipeds to raise their cheeping chicks in. We sketch the basic plan: a small rectangle of ¾-inch plywood for the base, four walls made of cedar plank, and two plywood roof sections, with a flared gable to overhang the perch stick. Ren, always the pragmatist, sees a problem. "How can we cut the boards?"

"No problem. We'll use a handsaw," I tell her. This is a perfect opportunity to show her that we do not actually need electricity. It is a luxury, not a necessity. I open the cabinet where my handsaws live in oiled splendor, their sharp toothy blades pointing inward, awaiting a moment like this. It's been a while since I used one.

"All right, let's try this one." It's an old 10-point crosscut saw, meaning it has nine teeth to the inch, perfect for cutting plywood and cedar. Not counting assorted docking saws, flooring saws, dovetail, coping, and offset dowel saws, I have four basic handsaws for nonelectric carpentry: a steel-handled 6-point for rough framing and cutting green timbers; my prize skewback 8-point, with rosewood handle and nail-cutting blade on the back side of the foreblade (sometimes called the "nib"); a 7-point ripsaw, with almost zero set to the teeth; and this sweet old 10-point crosscut, for fine work such as birdhouses. Time for a general lesson in handsaws, perhaps.

"Okay, honey—see how the back of this saw is thinner than the cutting edge? That's so it won't jam in the kerf. This is called 'taper ground.' Now, the way the teeth curve out is called the 'set.' Ripsaws don't need as much set because they cut with the grain, not against it like crosscut saws. Ripsaw teeth work like little chisels, but these crosscut sawteeth are knifelike." These were almost the exact words of my first boss, may the light shine upon him. Long ago, he showed me the V-groove on the cutting edge of his crosscut handsaw, and he slid a needle down its length, like a kayak going down a log flume, to demonstrate how evenly the teeth had to be set.

Ren says nothing, and I assume this means she is listening attentively. "Okay? Right. This is a 10-point saw. But when I talk about how many points per inch a saw has, it's always one more than the number of teeth per inch."

"Why?" Her trusting face, her innocent look—does it belie a cunning mind eager to trip her father up with hard questions? I recall that it took ten minutes and a diagram to explain it to me. "Honey, it's very complicated. Trust me. Anyway, lots of tiny teeth make a smooth cut, a few big teeth make a fast cut. And see how the handle doesn't have all the fancy 'fishtail' curves? That means it was made, oh, sometime after 1910, when they started streamlining everything—"

"Dad, are we going to paint this birdhouse?"

Screech, thumpity, crash, my train of thought derails. "Uhh . . . sure. I guess we can paint it." Sometimes I wonder about Serenity. She has benefited from a rural upbringing and tolerant parents, from cloth diapers and ecological coloring books, and more recently, a pony of her own. She seems to understand that our lifestyle is not simply a mindless poverty but a real attempt to live simply, thus avoiding a poverty of spirit. We like the smell of horses and chickens, fresh eggs for breakfast, and a shop full of wood shavings. I use a power saw for most jobs, but how do I pass on to her the ineffable rightness of a handsaw when I use one so seldom? For that matter, I also like the idea of writing in the ancient way, with a typewriter or pen. If I could, I would. My computer is more than a crutch.

Where was I? Something about . . . Yes. I'm bothered by the way Ren's mind won't stick to a single subject. Supposedly, it is either the hallmark of genius or of a disordered mind. It is as if she wants to absorb all the world's knowledge and beauty at once, feeling many ways about everything, intrigued by mystery and nuance: a trait author John Briggs called *omnivalence.* Wonder where she gets it.

Meantime, I'm laying out the various tools we'll need to assemble this birdhouse, once I cut the pieces: brace and bit, hand drill, screwdriver and plane, all non-power tools. "Here, sweetie, why don't you try cutting this piece of cedar? Smells nice, doesn't it? Just keep the saw straight, like so, and remember, a handsaw cuts on the push stroke." (Unlike all my fine-toothed Japanese saws, I do not say, trying to stifle my natural pedantic streak. They cut only on the pull stroke.)

Ren takes the saw, but she frowns. "I think electric saws are better," she decides. "You never use a handsaw, Dad."

Touché. However: "Yes, but handsaws are quieter, cheaper, and better for the environment than power saws."

Ren says nothing, although her eyes ask the big question: *Then why don't you use them instead?* She puts the teeth on the wood and draws the blade back. It must be in the blood; after a few tries, Ren gets the hang of it. As she works, finding the correct rhythm (zoo-ta, zoo-ta), she mentions a boy in her school, whose opinions tend to clash rather loudly with my own. His father is a logger. ". . . And like, Darrell says the environment is part of the economy."

I reel as if struck. "What? That's a steaming crock of . . . nonsense. Way off. That's like saying the horizon is part of the mountain." Although, by the time loggers are done with a mountain, it is. Time for a change of subject; all too soon, Darrell and his ilk will be on my doorstep, waiting for Serenity. Then I'll invite these boyfriends to pass the time by viewing my collection of horsewhips, knouts, cat-o'-nine-tails. . . .

"Good, that's a nice straight cut. Watch those splinters. That's right. The trick to using a crosscut handsaw is keeping it at forty-five degrees. With a ripsaw, you should increase the angle to sixty degrees by lifting the handle—wait, let me show you." I get a ripsaw, position a board on the sawhorse and make some sawdust, the old movements coming back to my shoulder like a memory. "See, just like this." All the pieces are nearly cut.

Serenity observes: "Power tools are a lot faster." I can't argue that. Nor can I quite explain the thrill it gave me, over twenty years ago, to watch my first mentor, Swanee, start his cut with a short, sharp push, slicing a neat groove alongside the pencil line. It was a lazy afternoon; we were packing up our tools, and he wanted to show me something magical. Like Ren, all I saw that day was an old man using an outdated tool, a bit of forgotten history. But he planted the seed in my memory, and it grew. It was poetry to watch him cut a board with a handsaw, stanzas of motion from another age when hand tools ruled the world of carpentry.

Only one more cut left. Overhead, the lights flicker, go on. Power's back. Ren looks at me. And suddenly, I have a choice to make.

Just as the twig is bent the tree's inclined.

ALEXANDER POPE

SLICK CHISEL

An Auldie but Guidie

Three feet long, this chisel lives in a brass-riveted scabbard made of Scotch-tanned bullhide. In the old days, it was a shipwright's and log cabin builder's tool. One doesn't tap it with a mallet; one pushes it, like a shuffleboard stick. The head is at least a hundred years old. Hercules could bend the handle without breaking it; I turned it out of rock maple and soaked it in castor oil for a month, polishing it off with boiled linseed oil. The steel takes hours to sharpen and holds an edge like a Damascus blade. It's a damn fine tool. I don't care what Freudians would make of it.

One night several years ago, a former acquaintance tried to buy my slick chisel, just before he became a former acquaintance. He was a metallurgist. "That's . . . really good steel," he said, thunking it with a fingernail. I handed him a small loupe—my shop is nothing if not well-stocked—so he could examine the recent razor edge I'd stropped on it.

After a minute, he muttered something about carbon and hardness. "Really good steel," he repeated, with an odd inflection to his voice. He offered a price; when I told him the tool wasn't for sale, he tripled it. He wanted, perhaps, to spirit it off to some nuclear laboratory a thousand miles away and bombard it with particles to force out its secrets. "It's not ordinary," he mumbled. "Pallasite, maybe. Unusual alloy of . . . Where did you find . . . ?"

I began to tell him about old Duncan, but realized midway into my narrative that the full story might overexcite a weak mind. Duncan was nearly ninety when I met him, a retired carpenter and a Scot to the bone. We got on well after I reeled off a few Highland proverbs to let him know I was not a complete ninny, and he gave me the head of my slick chisel as a gift. He swore it was no ordinary tool, hinting that it had been hammered from an old, old elven-steel claymore of fallen star-metal.

After I fitted on a handle, it was easy to envision Conan the Barbarian holding my chisel aloft on some smoking battlefield.

This guy didn't want to put my chisel down. "You don't know what you've got here," he said, not quite smiling. "You should sell it." To him, he meant.

I countered with a firm refusal. It's best to take a hard line with some people. They think the world is for sale, that any tool may be ripped from the bosom of its owner by chits of green paper. But—*Mony tynes the half mark whinger*, is my philosophy; *he that winna lout and lift a preen will ne'er be worth a groat.*

The virtue of such maxims, Duncan told me, is that they may be quoted to dolts who get in your face; when served with a snarling inflection, it squeezes the oatmeal right out of their bagpipes. Duncan had "hantle sich," meaning many such, and he shared a few of the best ones. They sound very profound and are completely unintelligible; you don't even have to worry about context.

No way would I sell this slick, ever. Besides, my biggest chisel would be wasted on a scientist. It has too much history.

The past is a wonderful time, especially one's own past: those younger dawns at twenty-one always appear in soft focus as if filmed through several layers of mental cheesecloth. It's possible to kid oneself on the subject of bygone days, and I may have a touch of that.

But I know this: Burdell Swanson was a genius in the craft. He used to tell me: "When you know all I know, you'll know it all." He was no egoist. He meant that I would never know all he knew, and he was right. His brain was a vast library of building lore, a subject so vast that I was still coming up with questions years after his death. Shortly after my mentor passed on, I felt the panic a fledgling eagle must know when nudged off the face of an escarpment. It was time to fly, but I didn't feel ready. My skills were intact, but incomplete.

For that reason, I joined a framing crew. In life, Swanee had scathingly described the type of degenerate who would wittingly become a framer: a side alley of human evolution, he insisted, whose specimens walked on their knuckles, mated with pickup trucks, and ate from the coolers of convenience stores.

My new co-workers were framing a number of giant apartment buildings in a little suburban cul-de-sac of a city I shall call Ecolopolis, on advice of counsel. I had worked on these structures before, but always in the very late stages of construction as a finish carpenter, a proud calling. Now I grunted and scratched daily with seven fellow framers, our hammers a blur and our minds a dial tone. I had been hired by a misty figure who appeared only on paydays to service the beer-lust of my cohorts. Amid the chaos, we built fast, if not well; no one seemed to be in charge, not even the nominal foreman. One tattooed crew member had learned his trade in prison, and another was ducking an indictment or awaiting trial; I was never clear which and was loath to ask.

All day long, the radio blared acid rock at apocalyptic decibels. At lunchtime, most of the crew disappeared into a silver trailer that was being used as a job shack. Fifteen minutes later, the door would open, expelling clouds of pungent smoke and coughing framers, blinking at the sunlight with bottomless pupils. They would stagger to a nearby supermarket and return with pastries devoid of any food value, jars of artichoke hearts, anchovy rolls in horehound syrup, and candied hummingbird tongues.

A confession: I spent one lunch hour in that trailer, mostly because I was curious to see what a cigar box full of Thai sticks looked like. It was not necessary to inhale; simply breathing the thick atmosphere in that bathyscaph addled me. Seven hundred years later, I fell out of the trailer behind the others and floated over to my last work station to try to recall how to don my toolbelt, whose complicated buckle I had never noticed before.

An hour later, the superintendent pulled up in his Land Rover and asked me how it was going, probably just to make conversation. I sputtered and radared the sky.

How was what going? The world monetary crisis looked bad. Personally, I found the going difficult, because now I understood the poetry of Robert Burns as if it were my mother tongue: *And aft your moss-traversing spunkies/Decoy the wight that late and drunk is: The bleezin, curst* . . . No, God, the question had too many ramifications. I found some spit and asked the laird to bide a wee while I sought out someone who could scale this tapmost towering height for him.

After that, I ate alone, and my stock descended.

These were men's men, used to altered states. After work on Fridays, they'd sluice down most of a case of beer before toddling home to practice their domestic backhand. On this crew, carpenters worked hard and took their pleasure forcibly. These dudes were capable of anything.

Heretofore, my heaviest hammer had been 22 ounces, but real framers used nothing less manly than 32 ounces, veritable sledges that sank framing nails in a single blow. (I bought one, prontissimo. It was way too heavy, at first.) My co-workers considered my best power saw a sweet little toy, but too puny for rough construction. They used worm-drive saws for everything, including fascia, siding, and corner trim. (I borrowed one. It also was too heavy, but I adapted.) Most damning of all, my hempy colleagues noticed I could function without a bloodstream full of drugs and alcohol, avoiding illegal refreshments on the lunch hour. (I pleaded allergies.) There were broad hints, delivered in semifriendly undertones, that I might not . . . survive as a framer. It was a stamina thing, I was told.

When completion of the third floor added serious altitude to our working conditions, the meaning became more literal, the warnings less subtle. But at some point or joint, I was conditionally accepted, or perhaps just periodically forgotten. I did my work and didn't let the side down, surviving by skill alone. For one thing, I could read blueprints as well as the nominal foreman; for another, I was the only worker who understood the relationship of the framing square to the wild hypotenuse. Even three stories up, my balance was excellent. I could walk across bare joists and trusses like a mountain goat, I never stepped on power cords that might suddenly be yanked, and all the early attempts to finesse me into a fall had failed. In those days I was filled to the brim with youthful sangfroid, fancying myself immortal. Only now do I realize how close to death every workday took me.

Came the Friday that we installed the roof trusses on the first building. On Monday, we would sheathe that roof and go on to the next building. Consequently, beer flowed after work. The conversation turned to one rafter that bulged above all the others; I had pulled the topic in that direction for my own ends. Everyone knew which rafter it was, and this teeny detail was the first job scheduled for next week.

"So take a saw up there and chop it off," the man with the tattoos suggested, to me or anyone else crazy enough to care. It was the usual cure for proud wood: snap a line and saw it off.

I went to my car and pulled out my giant slick chisel. All conversation stopped. Several voices invoked the Nazarene carpenter. Ignoring everybody, I climbed to the top floor and strode out onto the bare trusses to confront the bow, a blister of wood about a foot long caused by a knot in the rafter. I unsnapped the scabbard, placed the blade on the high spot and sliced it off with a few easy strokes, at once elated by the perfection of the cut and sad that it was over.

And then I surveyed the magnificent view of Ecolopolis spread out below, the bright sun at my back and every coin of my golden youth gathered up into one glistering pile of moments. My future was unclear, but I wouldna be a carl framer. *Ha! Wh'are ye gaun, ye crowlin ferlie?*

At the end of one perfect summer afternoon in 1975, I stood at the top of the world and the center of the universe, wielding this mighty blade before an audience of anthropoids who had tried to snuff or disable me for months. In that moment, I was top ape. There was drunken cheering below; for the rest of that job, I would be male-bonded to the max with these animals, my new pals. Triumphant, I let my eyes rest on the horizon of my life ahead, feeling the fullest truth of the old Scottish proverb: *When a man's gaun doon the brae, ilka ane gies him a jundie.*

I never drank the Muses' stank
Castalia's burn, an' a' that . . .

ROBERT BURNS

PIPE WRENCH

Of Steel and Lead

It has taken me five years to prepare myself psychologically to take on this old plumbing in the well-house. It looks to be a combat assault. Soon winter will descend, so it's either this morning or next year.

I hope our daughter learns to plumb someday, but I pray she never learns anything about steel pipe. Steel pipe is an instrument of torture, an invention of the devil. When I touch a pipe wrench to someone else's bad job of plumbing, my verbal work habits would burn out young ears. Today, while she's in school, I am attempting to correct an ancient and complicated mistake inside our old stone well-house. If she were home, I'd go broke in petty fines.

It will not be easy. The small door should be closed to contain the sound, because my wife is on the premises; but for working in claustrophobic quarters, I require both oxygen and daylight. My first task is to disconnect all the steel pipe from an antique holding tank and chuck it outside. After that, I'll part the Red Sea and turn a ping-pong ball inside-out without forming a cusp.

Once upon a time there was a moron. But he was a powerful moron, who could change the future merely by fouling up a job in the past, which was his present. In the late fifties, he took a pipe wrench and plumbed his house; now, in the nineties, it is my house. Having examined his handiwork, I have deduced a few things about this person. First of all, his IQ was lower than plant life. He was about three feet tall, yet his spine could bend into a perfect W, and his arms were at least five feet long. How else could he have tucked the water filter beyond that maze of pipes, where it could never be changed by anyone else? And he must have had X-ray vision like Superman; otherwise, why put the shutoff valve over in an inaccessible corner behind the pressure tank?

But first, he had to build the well-house itself. This was a massive achievement of bad engineering, his masterwork. He had to make the structure large enough to

RIDGID
HEAVY DUTY

contain the pump controls, water tank, and several hundred feet of misplaced steel pipe, but still small enough to preclude repair by anyone else. The headroom in this cold, dank, dark sentrybox of a well-house would be minimal, practically nonexistent for anyone taller than a leprechaun. Given the materials at hand, he chose rock, so that his misplumbed outbuilding would last forever. Finally, he designed the entire plumbing system like a ship in a bottle. *Go ahead, sport,* his ghost seems to whisper. *Crawl in here and fix something.*

I have done so, many times, wishing I knew his name or just the location of his grave. The shutoff valve is on the wall farthest from the door; to work on the indoor plumbing, I have to enter the well-house, climb over shin-murdering pipes, slither into a narrow gap between the tank and the wall, stretch around the holding tank to the farthermost limits of my fingertips, and turn off the valve, which leaks, of course. It is a vile job even in the daylight, but at night it simply cannot be done. All plumbing leaks occur at night.

This nameless individual plumbed the entire house, but over the last five years I have effaced most of his original plumbing from the universe. My tools were a hacksaw, a vocabulary, and a pipe wrench. Miles of old steel pipe crisscrossed under the floor joists, in areas that spiders would consider cramped. I ripped it all out, including one trombone from hell that used twenty feet of elbows and nipples to go five linear feet, sawed it into lengths that would fit up through the crawlway access hole—a funless task, in the dark and wet—and replaced everything with copper and ABS, until my voice was hoarse.

As a child, I did not have any great desire to be a plumber, although in retrospect, it might have been a wise career direction. But I watched my father plumb a number of the houses we lived in. It was his least favorite hobby, and he always muttered nonstop to himself: "Who did this? What sack of . . . *idiot* would . . . oh, you miserable, dirty, godforsaken, syphilitic . . . ouch . . ." His karma was such that those who plumbed before him seldom installed shutoff valves of any kind, and he was forced to do this awful work on weekends, usually Sunday nights when the hardware stores were closed. Because of this, his naturally inventive streak was fueled by the mean mother of invention, necessity. He could wrap a leaking pipe with a bicycle innertube so well that it would last for years. He's gone now, but I can still see his lower torso sticking out of a sink cabinet, legs kicking and water spraying in all directions, unclogging his larynx with amazing new words.

After high school, I took a summer job at a college, and there learned all I ever want to know about steel pipe. A spotless boiler room was the hub of the Building and Grounds department, and Jim, the red-eyed chief engineer, kept everything ticking. He was built along the lines of his boiler, huge and hot, and he taught me all sorts of things, including how to cut pipe. Galvanized steel pipe came in twenty-foot lengths that could be cut and threaded with a pipe vise, a reamer, a stock and die for cutting threads at both ends, a pair of mighty arms, and two pipe wrenches laid nearby. These days, most plumbing supply stores have a handy machine for cutting and threading steel pipe; but in those days, in that place, we did it all by hand.

There was something vaguely pleasurable about the process, creating sturdy tubes to transport hot water from the boiler to every building on campus. One had to add an extra two inches on every cut, to allow for the threads. The cut had to be perfectly square, and the burrs reamed out afterward; any little pieces of metal inside the pipe would collect sediment, eventually plugging the pipe. "If you cut the threads right," the chief claimed, "you can turn the pipe three full turns into a fitting before you need the pipe wrench." Jim always counted them aloud, smiling like a deranged Mister Rogers turning the winch on a torture rack. We used plenty of cutting oil, so the sharp dies needed only a strong arm and a good eye. Since pipe threads are slightly tapered, the deeper they go into the fitting, the tighter the joint becomes. Jim also showed me how to smear pipe dope (pipe joint compound) over the external male threads, but never on female fittings. "That'll plug up the pipe someday," he explained.

I loved those happy days. If they didn't turn me into a plumber, nothing would. Nothing ever did.

Without pipe wrenches, you can only stare at steel pipe. No other tool will budge it. When set to the proper diameter of the pipe and fitting, pipe wrench jaws are self-tightening, but only in one direction. To tighten an elbow onto a pipe, you need two wrenches: one to turn the elbow clockwise, and the other to hold the pipe in place by the application of brute force in a counterclockwise direction. Without that second wrench, you might put enough strain on the pipe to turn something else down the line. The trick is to balance the opposing forces in such a way that only the elbow turns and the pipe stays absolutely still. It helps to keep the two wrenches within four inches of each other, one turning, one holding.

When buying a pipe wrench or three, go for the best. Look for brand names. I once broke a cheapo pipe wrench made far across the sea, and it failed at an awkward time, which made for a most unpleasant day. Now I own five of the best in varying lengths, and keep the knurled nuts oiled so that I can easily adjust the jaws with a few twirls. Pipe wrenches do most of their work around water, so be sure to wipe and oil them, especially the moving parts, before putting them away in a dry place. Red or orange colors make them easy to locate, but rust brown just fades into the background.

Steel pipe may seem obsolete in these days of soldered copper, but there is still plenty of it around, and it's strong enough to stand on, kick, or trip over. Up until the last century, pipe joints had to be made by hand with hot lead; hence the name "plumber," from the Latin word for lead: *plumbum.* Also see "lead poisoning," or *plumbism.* Now we know that lead and drinking water don't mix, but it seemed like a perfectly good material to the Romans. That's how long it took civilization to discover that lead is cumulatively poisonous, even the tiny amount of alloyed lead in the soldered joints of old copper water pipes. Code states that you must use a special solder—one that is lead-free—to connect them.

If you have a taste for masochism, you can recycle steel pipe. In only a few hours, I've managed to remove an entire elbow and three feet of pipe. The holding tank is connected to the main pipe subsystem by one last union, which is frozen solid. Nameless did not plumb well but permanently, and his pipe compound was some kind of epoxy cement. Usually I am able to use a cheater bar, a length of 1½-inch steel pipe that slides over the pipe wrench handle for extra leverage to unscrew fittings. But in here, there's no room. I could also play the flame from a propane torch on the joint, which expands the female threads enough to loosen them. The basic idea is good, but it has already cost me one eyebrow this morning.

Putting both hands on the end of the wrench handle, I strain to move it by force alone, grunting a little litany of curses.

Eventually, the pipe relents, turning a fraction. I get a better purchase on the pipe and push. Another tiny movement counterclockwise. Success. With luck, I'll be done tomorrow.

General Taylor never surrenders.

THOMAS L. CRITTENDEN

BLOWTORCH

Quest for Fire

And here it is, the moment of truth: my disposable propane bottle is empty. It feels light, unlike the satisfying spongy heft it had when I first bought it. It has lasted three months, more or less, in occasional use, which is just about right. Several upcoming projects will require a torch. And I have a fresh replacement tank right here on my shop shelf. How about that.

I could—that is, I *could* use one of my old brass blowtorches. It might be a good idea, very ecologically sound, and perhaps only slightly more dangerous. Certainly it bears thinking about.

Slowly I unscrew the torch head from the empty propane tank, visualizing all the possible uses of this disposable empty metal cannister. They float; with fifty of them, I could build, say, a swimming dock for otters. But there's only this one. Yonder is the garbage can. Why not? That's what one is supposed to do with these propane tanks when empty.

How did we come to this? At one time, the word disposable had no real meaning. Things were not designed to be thrown away after use, even containers. This was in the golden days before we all became consumers: another strange word from nowhere.

During the seventies, I shaved with a straight razor. It was entertaining, time-consuming, somewhat perilous, and very educational. One day as the eighties began, I used a disposable razor with a plastic handle, and something awful happened inside my soul: I liked it. The idea of disposability was abhorrent, of course, and at first I felt like Richard Nixon scraping his dark jowls in the morning. But the blade was perfectly sharp with no effort on my part, and I could pretend that I was going to use it over and over again, until it wore out or I lost it. This razor lasted a week, and then I lost it in the vicinity of the garbage sack and bought another. Like drugs, sharp razors quickly become an addiction.

Blowtorch

For a full year during the Me Decade, I used disposable razors, pens, and lighters without any guilt or consciousness at all. In 1983 I came to, and began weeding out the disposables. It was a shameless time in my life, never to be repeated, but it was a wonderful lesson in tolerance for people who sometimes use disposable products, including yours truly.

Before the propane torch, the blowtorch was the right tool for applying flame to a workpiece. Plumbers always had one, and they used it for sweating copper joints, thawing frozen pipes, and melting lead. Electricians used them for soldering wires. Painters used them also, for softening window putty or removing alligatored paint (not a safe practice). Machinists employed them, and tinkered with them, and for that matter, it was machinists who designed them in the first place. No one threw them away. They were everywhere. But today, you'll look long and hard to find one.

I have two, oldies in beautiful condition salvaged from different garage sales. Maybe it's high time to start using them both. They'll run forever on white gas. It's a little more work, to be sure, but no detritus remains to be discarded in a landfill after the job. A blowtorch produces only heat, not waste products.

The only snag is, I don't know how to make them run. In fact, they have never been used since the day I found them at a yard sale. Perhaps they don't work at all.

I was nineteen years old the first time I held a blowtorch. Swanee had a pair of them, which he filled up one morning with white gas, pumped, primed somehow, and lit with a wooden match. I took one, roaring in my hands like a shiny newborn dragon. If memory serves, we were heating rusted steel pipe joints to dissolve the corrosion so that we could take them apart.

Recalling those lost days, I fetch the blowtorch from its shelf in the shop and bring it inside to the kitchen table. With its bright red handle and shiny fat brass body, this is really a beautiful tool. I pump it up. Holding a lit match to the nozzle of my torch fails to have any effect, although a little gas runs down the side and almost sets fire to the kitchen table. My wife begs me to take it outside to light it. Half an hour of pumping and tinkering with it on the patio fails to result in ignition. There are three mysterious adjusting screws, and I play with each in turn. Nothing happens. Truthfully, I'm afraid to pump up the canister too much.

The blowtorch is so obsolete that no book in our local public library has any detailed information on it. There are a few tantalizing pictures of a blowtorch in operation, and even a couple of references in some very old carpentry and plumbing manuals, but the basic assumption seems to be that one is using a working blowtorch, not a broken one, or that the reader knows how to light the darn thing. In fact, these may be the only words written on the subject of blowtorches in the last twenty years. An online database search also yields zip.

By asking every person who might know, I finally discover one individual, only recently deceased, who knew a lot about blowtorches. He can be reached by Ouija board, Dead.com.@grave.forestlawn. His coworkers give me the number of another old expert, one Walter, on whose answering machine I leave a message: I'm trying to find anyone who knows anything about repairing and using blowtorches. Can he help?

Propane torches began to proliferate in the early sixties, inevitably supplanting the smelly old blowtorch. The original, small (1–3 lbs.) propane tanks were refillable, and you can still buy a professional plumber's rig with refillable tank for about $300, which would buy a lifetime supply of convenient nonrecyclable propane cyclinders. Mapp gas is available in disposable tanks, and it will give you a flame that is almost 500 degrees hotter than propane. Why, the world seems to ask, would you want to use a cruddy old blowtorch? The name lasted longer than the tool; oxyacetylene torches sometimes were called "blowtorches," but that nomenclature didn't last.

There does not seem to be a lot of current blowtorch lore; apparently no one is using them in the present day. One would almost think the blowtorch was a relic of some ancient yore, like crank telephones.

However, there is rather a lot of data available on propane torches, especially in homeowner's and how-to magazines. Nothing could be simpler to use: Screw the brass burner pipe unit onto a disposable propane tank, turn the knob slightly counterclockwise, and ignite the hissing gas with a match or striker. The cone of fire can be adjusted with the knob, and will produce enough heat to melt solder, unfreeze rusted pipe joints, peel paint (still not recommended—fire hazard); generally, the propane torch will do all the jobs a blowtorch used to do. It's hotter, for one thing, and the flame is smaller, more controllable.

A few years ago, my wife, Joy, snaffled my old propane torch and began experimenting with laboratory glass. Shortly thereafter, she took out a business loan and purchased a full oxypropane rig, a remarkable tool in that I am not allowed to touch it at all. At this writing, Joy is selling her line of handmade glass jewelry in outlets all over the United States and one in Zurich, Switzerland. She uses discardable cylinders for some phases of production, and no, she has no problem with the philosophical ramifications of discarding four or five empty propane tanks a year. A blowtorch, she says, would be of absolutely no use to her whatsoever. "All right, it's prettier. But is it as hot as a propane torch? As clean? As convenient? As safe? Nope? Nope. It's cute, honey. Take it away."

The newer propane torches feature a neat piezoelectric self-igniter, which creates a spark using roughly the same principle as crunching up wintergreen Lifesavers in the dark (try it if you're curious, but don't use your own teeth).

It would be glorious to light my own blowtorch with a wooden match. I would derive a lot of personal satisfaction from rebuilding this obsolete tool and making it function once more.

Walter calls. "Basically, it's an antique," he says. "I mean, you could use one, theoretically, but it's not really as safe as propane." Unsafe? In what way? Walter amplifies, relating a few old stories spiced with phrases like *whump, engulfed in a ten-foot orange fireball,* and *human torch.* I get the idea.

"Thing is, there's no way you should light one indoors, especially if you're having trouble getting that puppy lit. You're not trying to light that sucker on your kitchen table, are you? If I were you," Walter says, "I'd just put a nice shine on it and use it for a paperweight."

The blowtorch has a classic beauty that time cannot mar. It evokes a simpler, better world when nothing was ever wasted. It makes a grand paperweight, stable and elegant, far superior to any propane torch.

CROWBAR

The New Barbarians

Washington, D.C., looked like a scene from the French Revolution; hasty scaffolds were being set up in front of the Capitol, and shivering bureaucrats stood in the rain on the beds of pickup truck tumbrels, stripped of their neckties and awaiting their turn at elevation. The halls of power were crammed with rioters; over on Pennsylvania Avenue, the Oval Office doors finally splintered and fell. Water dripped off the end of Phil's long nose as he faced the cringing President of the United States, crowbar held at port arms. His pointed beard made him look like the Devil, come to collect.

". . . And I'd say, 'Mr. President, you are under arrest. In the Name of the Working Class!'" He brought his crowbar down in a two-handed smash, burying the crook in the presidential desk, played by an old sawhorse. Phil snarled, "No more perks, no more waste, no more taxes, no more corruption. We'd stick their heads in the toilet and flush. Then we'd take over." He looked up at me. "You with us?"

I was standing on a ladder, listening to Phil rant while I nailed wet drywall backers to the wet interior walls of what would be a dry and comfortable one-of-a-kind custom house in a tract subdivision, Oxymoron Estates. It was hard to be enthusiastic. Someone ought to lop off the Washington Monument, maybe, but I wasn't up to it. All I wanted was some sunshine. The rain still fell half-heartedly, not like the real drenching we'd received an hour before, stopping all work. Payday was next week. You say you want a revolution?

Phil looked and talked like a young Trotsky, if Trotsky had a blond goatee and had taken lots of methamphetamines, although Phil had no truck with powders: "True revolutionaries shun drugs, except for coffee and cigarettes," he claimed. The owners of this house would never know the sedition that blossomed under their roof, which was in the process of being sheathed with plywood. Ferocious hammering over our heads sounded like cannon fire.

There are two ways of looking at a crowbar. It's a tool for construction and destruction; structure is the root of both. "Naw, count me out, Phil." Bloody revolutions don't change anything, except for fresh tyrants. "I'm hoping for divine intervention, like an earthquake or something."

He wrenched the crowbar loose from the trestle. "Yeah, well, when they come for you some night, you'll change your mind."

They had come for Phil recently. A cop had pulled him over for a missing taillight, found a warrant on file for an unpaid fishing violation, and hauled him off to jail in handcuffs. He missed work a few days and came back a new man, committed to a course ever since his arrest. Knowing Phil, I suspected he did not go peacefully.

Nor was spring coming quietly to Oregon. The weather gods had ordained an extra month of squalls and vertical liquid, typical of the Pacific Northwest but a real shock to the refugees from California who were flooding north in search of a new beginning. Phil was such a transplant, a black-and-sweet coffee addict who drank enough to qualify for Turkish citizenship. With caffeine and attitude, he fought the mood swings of Seasonal Affective Disorder, a sky-induced syndrome that makes suicide look like blessed relief. Rage was his stimulant of choice, directed at hegemony of any kind; he kept his anger well fed by reading newspapers and underground screeds. He was the first real anarchist I ever met, with a degree in political science and a hunger for chaos.

He was also a student of the crowbar, sometimes called a wrecking bar because it works so well for demolition. When taking down buildings, you want a lever that won't bend or break. But the crow is an underrated building tool as well, a steel stick that will lift a frame wall off a deck or turn a warped joist for nailing. Phil loved the way his bar pulled nails, and talked about moving railroad boxcars with a six-foot crowbar under one wheel. I didn't completely believe him until I saw it done, years later. Archimedes was right: with a big enough lever, you can move worlds.

The original blacksmith's name for it was "crow," God knows why, and it goes by a few other names today. The pinch bar, a crookless bar that can be up to six feet long, works beautifully for stripping down concrete forms or tightening wooden braces. A digging bar is little more than a long iron prod with a pointed end, but it's ideal for breaking up rocks when the post auger hits gravel.

The most common bar, especially for finish work, is the flatbar. With one of those, you can adjust door jambs, lever heavy doors into position while you replace the hinge pins, pull small nails, rip off cove molding without damaging it, raise one side of a window to level it, and countless other jobs. If I could only have one bar for light construction, it would be the flatbar. (But I can have as many as I want. They're cheap.) Got a joist hanger in the wrong place? Use a flatbar if you'd like it taken off intact, or hook it with a crowbar if you just want speedy removal.

The flatbar shines for general prying, but for raising the edge of asphalt shingles to slide in a repair shingle, it has no equal. The prime time to do this, oddly enough, is when it's raining, preferably a freezing rain that soaks your behind as you sit there on a roof prying up shingles with numb fingers. The shingles will rip on a sunny day. Don't try to pry up two tabs at once; use both hands and your full attention to lift up the edge of a 3–tab shingle, one tab at a time. Speed causes rips, which in turn will eventually cause leaks.

You can use the flatbar to remove roofing nails, even when the head is hidden under the shingle above them. And come to think of it, the flatbar would be a superlative tool for loosening gallows trapdoors, if they swelled tight in the rain. (A well-built trapdoor is beveled on the top to avoid this, but who has the experience to knock up a multipassenger gallows these days?)

Phil and I went our ways, but a couple of years later we met again when I was hired to help tear down a barn for a gentleman rancher, a corpulent Clint Eastwood look-alike in Stetson and cowboy boots. He aimed at the barn with the slimy tip of a hand-twisted cigar. "You'll be working with that fella. I sure wouldn't piss him off." There by the barn was Phil, already setting up a ladder.

At first, he didn't look entirely happy to see me.

"Look, don't say anything to the boss about . . . you know, what we talked about 'way back when.' He's a real filthy Republican, thinks John Wayne should be the pope."

"The dirty swine," I said, watching the owner's shiny pickup gallop down the driveway. "Okay, we can scream now. *Death to the ruling classes!!* Will he be one of those we hang? Comes the Day, I mean?"

"Right at the head of the line," Phil said, smacking his palm with the flat of his hammer. "Bastard is making two hundred grand a year, but he gets muley over a few dollars an hour." Apparently Phil was outraged when he heard the first wage offered. But the cowboy called back, Phil said, after he found out that five bucks an hour wouldn't hire anyone but drunks, stoners, students, and illegal immigrants; "You know, the usual junkies and parasites." Phil drank like a French sailor, but he was always sober during working hours.

This was a good thing, because taking down a barn ranks right up there for easy access to death, dismemberment, and paraplegia. We set up scaffolds and rigged safety lines; the first thing we salvaged was an antique weathervane at the flared end of the gable, easily forty feet off the manure-soaked ground. "Fall off on this side, and the EMT's won't let you ride in their meatwagon," Phil grunted, balancing on the verge over the muck. He would have liked to have torn down the whole structure with a Cat and chain, and burn it all, but the rampant capitalist owner reckoned to save the old-growth beams that ran through the body of the barn. In addition, every board had to be stacked and every nail collected in a bucket; horses would be pastured in this field next year. He allowed four weeks to finish the job.

The first two days, we used spades and flatbars to strip the cedar shingles, moving down the gambrel in widening arcs that marked the ends of our safety lines. They made an excellent bonfire; Phil pointed out to Tex, as we called him by then, that the shingles were burnable garbage, and a magnet would pick up all the roofing nails from the ashes. Then we had to take off the sheathing boards, one at a time.

We were desheathing the roof by the third day, letting the 1 x 4 boards slide down to the dirt. At lunchtime, Phil took a closer look at my crow, a well-made bar with flecks of the original dark blue paint remaining. "Good bar, man. What's that stamped into it? Looks like four initials."

I held it up so he could read them: USAF.

Phil grabbed his face with one hand, eyes bulging between his fingers. He looked like an elephant passing coconuts. "Jesus God! That's a $7,000 crow bar! It's rated for polar warfare and underseas nail-pulling! That's the finest carbon-steel known to man, according to the contractor bastards who sold it to our great-grandchildren! Where'd you find it?"

"Five bucks at a garage sale in Colorado Springs."

"Goddamn the government. What the hell does the Air Force need a crowbar for?"

I had asked the same question of the man who sold it to me, and got an answer: for prying canopies off burning aircraft. It was a standard steel bar, made of harder metal than the average crow, and I have to admit that of all the wrecking bars I have ever used, it's still my favorite. It is 24 inches long, octagonal, with a sharp nail-claw at one end and a slight offset at the pinch end, for leverage. My other big crow is a full 3 feet, and its gooseneck crook has a special purpose. Slip the crook under a joist on edge, and you can turn the whole board against the header joist; this is an excellent way to true up warped joists.

My smallest crow is a burglar's jimmy, the one Indiana Jones uses to pry up crate lids, a tiny civilian version of my Air Force crowbar. In some states, mere possession of this useful bar is probable cause for arrest; it works better than the key on a locked door. Phil always carried a jimmy in his truck, on general principles. "They can't tell me what tools are legal," he said. "I'll carry one if I damn well feel like it."

At the end of the job, just as we were packing up, Tex drove out to the site. "Mighty good job, boys." He looked at the neatly stacked boards, the immaculate, nail-free grass. Then he reached in his pocket and pulled out two checks. "It wasn't easy to do it right, neither. There's a two-hundred-dollar bonus in there for each of you."

It was totally unexpected. I didn't exactly fight back tears, but I was touched. And I know what you're thinking: Phil would be struck by a deeper understanding of human nature and realize that not all hats were black, that only a few capitalists were inhuman oppressors, and go on to pursue nonviolent political solutions with the Green Party.

Not Phil. Not likely. He took the check with a grunt that might have been thanks and got in his truck. End of story.

But I dream about him sometimes, apocalyptic dreams of fire and carnage, marble toilet stalls with feet sticking out of the bowls. He's always holding that wrecking bar, usually smashing down something unconstitutional. Somewhere out there, he's waiting for a crack in the system, just enough to slide in the tip of the pinch and pry it apart. Comes the day, he'll bring Shebang, D.C., down, if he can.

I won't show up. May it all collapse from its own moral decay long before that, so we can rebuild something better. I believe in Crowbars for Peace.

The urge to destroy is really a creative urge.

MIKHAIL BAKUNIN

DRAWKNIFE

Estate of Grace

Only one thing captures the piquancy of knowing something is damnably wrong and being unable to stop it, and that's knowing you are a tiny part of the problem. Log trucks pass our house every day, some carrying a single giant log older than anything but the sky and the earth, certainly older than the Industrial Revolution that eventually killed it. In our little valley, much of the surrounding forest has been clearcut. I would prefer the view, if anyone asked.

But wood comes from trees, as someone once wrote. Steady demand creates a need for supply. As a carpenter and writer, I demand that wood and paper be immediately available, and reasonably priced. Free would not be an unreasonable price.

That's what I keep in mind when I use dimensional lumber. It's a small coin of guilt that doesn't change anything, but it keeps me from wasting any kind of wood; that includes the rough wooden poles, debarked with my drawknife, that I'll use to build a barn for our daughter's pony.

This may not sound credible, but I believe Serenity may be the current incarnation of Her Serene Highness Princess Grace of Monaco, who died about the time Ren was born. I can't prove it. It was hardly remarkable when Ren named her dolls Stephanie and Caroline. When she was four, we were driving on a twisty old mountain road, and she worried that the brakes might fail; I thought that was amusing and nothing more. In itself, the way she glued herself to the TV whenever it mentioned Monaco or the Rainiers meant little. But one night, she said something in her sleep—in French. And a light in my head went, *Click.*

A few days later, I asked what she thought Philadelphia was like. She closed her eyes: "Red bricks, lace curtains, old elms." Grace was raised in Philadelphia. Serenity has never been there.

Anyway, now we need a barn for her pony. It happened like this: when she was nine, Ren spent the winter wondering—often and aloud—when her apparently inadequate parents were going to buy her a pony. This year? Next? Ever? She had done the initial work: learning to mount, ride, saddle, and hoof-pick a friend's horse. And every person in her school, including the teachers, owned a horse except her, the lone pedestrian in the third grade. We were able to disprove most of this last, but to no real avail. "Some girls are, some girls aren't. But if Ren's horse-crazy," our neighbor lady explained, "it'll just get worse."

Horse-crazy? What's the etiology and prognosis of this insanity? No one knows. But it's incurable, and really expensive. Grace Kelly learned to ride at a tony boarding school. Sure, her father was a rich contractor.

We tried reason. "Look, honey," I said. "We'd like nothing more than to buy you a pony. Here's the problem." I ticked the points off on my fingers: "First, we need a pasture. The neighbors have one but they'll no doubt want money for it. We also need a barn. Not to mention saddles, tack, feed, vet bills, etc. And even after all that, we have to find the right pony."

Ren gave it a moment's thought. Nine years is not the age of reason. At nine they are all little lawyers, forever negotiating terms with their parents, gaining new territory and always adding to it. "I know where a pony-for-sale is," she said helpfully. But we had already checked on that pony. He was spoken for. We figured we were off the hook, for a while.

One night, the pony owners called. The people who wanted him for their infant had decided to wait a few more years. Would we consider keeping the pony? He came with saddles and tack, of course, and they'd chip in on vet bills if we took care of the maintenance . . . .

I put down the phone. The salient aspect of Grace Kelly's life was her optimism that things would all work out. For the cost of oats and hay, Serenity would have a horse of her own. Of course, we'd have to build a barn now.

We still had no pasture. The adjacent farm had been in the same family for the last forty years, but just as suddenly as the phone call, it went up for sale and was quickly purchased by my brother and his wife. "You guys ought to have the pasture from your house down to the river," they said, "now that Ren is getting a pony." (The miracle of telephones: Serenity had already talked with her aunt and uncle.)

So a barn was inevitable. But according to the chief budgeteer, I need free materials. "How about poles?" Joy asks.

At first, I resist the idea of building with poles. To begin with, I'd have to gather the fresh unprocessed raw lumber from one of the clearcut moonscapes around our home. I appreciate the concept of such gleaning, but not the massive harvest that makes it possible.

Even straight poles vary. They're not uniform. Nails won't work; I'd have to use spikes and pegs, notching my joins and approximating all dimensions. The few times I worked with poles was in a brief Peter Pan stage, when I wore a large belt knife with my buckskins and beads. I was younger than Ren.

On the other hand, the materials would be free for the hauling. I have a pickup truck.

The woods were lovely, dark, and profitable. Entire hillsides are now fields of stubble, but there are lots of usable poles on slash piles, the residue of a modern wood harvest. The loggers didn't vacuum up everything.

Gathering the poles brings back pleasant old memories, including a few of past lives in simple hunter-gatherer societies. I load several dozen of varying diameters on my motorized travois and go home.

For over twenty years, I've built in dimensional lumber, as it's called—four sides and two ends—or used peeled factory logs of roughly uniform thickness to build log houses. I like the precision and angular lines of a square, plumb, straight structure; when the last nail is set, something in my left brain smiles a square, straight smile, exactly plumb with the end of my nose.

But now my right brain is asking for a little creativity in the matter. That's reasonable. The pony won't mind a bit if his house lacks miter cuts and perfectly square inside corners. He won't care if it's a little rustic. And it's been too long since I worked with my drawknife, bitbrace, and chisels.

I learned carpentry from a number of teachers, including a Dane, a German, a Scot, one badly disoriented Irishman, and a Native American whose ancestors built exclusively with pole stock, not very far from here. Black Elk, the Oglala holy man, said there was no power in a square, but if you look at the average boxy edifice housing a powerful multinational bank, that mystical pronouncement seems obviously false. *Unless he meant another kind of power*, my right brain chimes in.

So it should make old Half-a-Brain happy that I'm building the barn out of poles. In working with poles, you can be off an inch and it will not show to the keenest eye.

For rinding (debarking) my poles, I set them on a sawbuck, turn the beveled edge of the drawknife down and shave off long strips of bark, turning the pole as I go. A bark spud would work better on larger logs, but for poles, a drawknife is the way to go. Lincoln might have used one of these, back when he split rails for fences. Maybe not. But it's a good tool for making big tenons.

It takes two entire days to strip off the bark, auger and clamshell the holes for the uprights, set them, and put on the lintels, using simple mortise-and-tenon joints. I use my drawknife every step of the way. It's particularly suited to making notches for the braces. The blade is slightly convex, curving away from the handles; the edge is beveled on one side, and different results can be obtained by turning the bevel up or down on the pull stroke.

Perhaps the drawknife came to carpentry by way of the chairmaker, who used it to form the chair legs, stretchers, and other turned parts before they went to the lathe. But every branch of woodworking had its own type of drawknife, with innumerable variants. An old tool, it ranges in size from 8 to 18 inches; the Vikings used them earlier than a thousand years ago to smooth surfaces after adzing, and they came to America with the earliest European settlers. Coopers modified the original design and developed specialty drawknives to shape staves or cut the bevel on the cask head. Shipwrights used them to make masts for clipper ships, wheelwrights for shaping spokes and shafts or trimming felloes, handlemakers and gunsmiths for paring off waste wood on rake and shovel handles or gunstocks. Woodcarvers employ miniature drawknives, 4 inches or so, with tear-drop handles.

By the end of the week, the barn is standing. As always with any experiment in carpentry, I'm a little surprised that I knew how to do it. It evolved without a blueprint, and yet the plan was always in my head. The joinery—mortise and tenon for the bracers, crosslaps for the main members—was adapted from dimensional carpentry. I give it a shake and nothing moves. It's sturdy.

And very rough. Now some part of my *right* brain smiles lopsidedly.

It was a very small barn for a very small pony, and thus a small barn raising, involving only myself, Joy, and Ren. Someday when I have a few million in cash lying fallow, I intend to design and construct a vast barn out of dimensional lumber. No expense will be spared. I'll store my sport blimp therein, perhaps a helicopter over in one corner.

But until then, this lilliputian structure brings a certain country ambience to Oxtayle Acres. A white and miraculous pony, half Arab and half Welsh, now calls it home. His name is Silver. Ren hops in the saddle with an easy grace and gallops off into the sunset as far as the fence. She reins in her little steed; the brakes work fine.

The best way to make children good is to make them happy.

OSCAR WILDE

CHISEL

Keen Edge, Steady Hand

It's a terrible thing to lose one's fine old tools. ". . . And all my chisels," Bill moans. "Man, some of them babies were eighty years old!" I feel his pain; Bill is a contractor, a fellow carpenter, temporarily devastated by a fire on a job site that consumed all his heirloom hand tools. "Ouch," is all I can say. It makes me queasy just to think about tools burned beyond recognition.

This Saturday morning, Bill and I are members of an eight-person work party gathered outside the local community center. It is a huge old hall built in 1926 without a foundation, by otherwise expert carpenters back in the otherwise good old days. Those old boys had a town to build around this roaring logging camp, with saloons, churches, bawdy houses, general stores, and no time to waste. This is what being in a hurry does: by 1995, one corner of the building has sagged down into the dirt.

The eight of us make up two centuries of combined carpentry experience, including timber framing, dome building, and barn raising. We slither in under the building, jack up the floor joists on one side, and remove the damaged section of beam. Sixteen feet of 10 x 10 fir is now light enough for one person to carry to the burn pile with one hand. That's dry rot.

"Watch that jack," Al warns, his entire body under the main beam. "I had great sex last night, and I want to do it again tonight." There's an *ad hominem* argument for you. Makes sense to the rest of us, to have such a good reason to live. He gets a few offers to fill in if he can't make it, the usual coarse talk.

Finally, we get all the rotted wood pulled out. All right, the new beam is ready to go in, but we've got to notch the end where it crosslaps at the corner. Now comes chisel time. We all have them, in great variety. (Except for Bill, that is.) Our toolchests are positively crammed with chisels: firmer, paring, socket and mortise chisels, big slicks, little carving gouges, matched sets and singletons. We spend some time comparing

them before we start cutting the beam with big mill chisels, the same kind of tool that made the original cuts so long ago.

There is something in the coveting of a strange chisel that is both satisfying and disturbing. You take it reverently in your hands, if the owner even lets you touch it, and you give it back as soon as possible. Nobody loans out sharp chisels overnight. When they are for sale, chisels go quickly unless they are priced beyond the pale of sanity, as is the case with most antiques. A new chisel is a cool delight: sharp-edged, untouched, virginal. But an experienced chisel is somehow warmer, friendlier to the touch.

I have an old Stanley ½-inch paring chisel, part of a set long scattered to the four winds. (N.B.—When you come across one good chisel at a sale, estate, or farm auction, look around for the others in the set. Even in the hands of many owners, a set of matching chisels will tend to stay in a general geographic area for decades.) After fifteen minutes of sharpening, this chisel will stay keen for several weeks of moderate use. I wouldn't dream of bashing it with a mallet; it is exclusively used for semi-fine work, pushed with the heel of a hand.

My tool for work like this is less delicate, a beater firmer chisel with a polycarbonate handle and a 4-inch blade, of a type that is sometimes called a pocket chisel, which refers to its length (6 to 8 inches) and to where it lives: it lives in a pocket of my toolpouch, and serves for most rough work. It was designed to be struck—brace yourself—with a hammer, although that's carrying heavy-duty a bit far. Remodelers often keep one of these expendables for general use, or else a sturdy mortise chisel (larger, wood-handled, with an overall length upwards of a foot.) The blades are often quite short from repeated grinding after hitting nails.

Chisels go back a long way, to neolithic times. The Romans refined the firmer chisel (from *fermoir*, to grasp; early chisel handles clamped onto flat blades) and the two basic styles of chisel construction, tang and socket. A socket chisel blade is shaped like a hollow cone, into which the handle fits. Tanged chisels have a tapering point that is driven into a hole in the handle; on superior chisels, like my old Stanley, the tang goes completely through the handle and terminates in an end cap. Handles on firmer chisels are made from beech or ash, sometimes octagonal, and ringed with a steel hoop at the end to prevent mushrooming from repeated mallet blows.

Paring chisels are the tools of cabinetmakers and joiners, and the sides of the blade are beveled as well, so the sharp corners of the blade edge can clean out the snuggest corners of mortises, dadoes, and gains. With boxwood handles and a shallow angle (less than twenty-five degrees) to its cutting edge, paring chisels are usually tanged rather than socketed, since they are pushed with the hand like a plane.

Saturday passes in a day-long orgy of community labor. By sunset, the grange hall is jacked up level, with new treated posts resting on concrete piers. We examine a crumbling temperance poster, circa 1926, proclaiming that *Saloons Raise Taxes!*, a sentiment we all toast in beer. We write our names on a piece of paper as a time capsule and staple it well up under the eave line. Another seventy-five years from now, the next crew will know who did the work.

The next morning, my body feels like a corpse in repose, every muscle groaning and joints locked. I mull over some possibilities: Tetanus? Premature old age? Fortunately, it's only the flu, whose symptoms are compounded by sore muscles. This strain of influenza traveled all the way from Asia to get me, and the only medical treatment for these viral infections is rest, fluids, television, and gallons of chicken soup or brandy. I have fevered dreams of all my chisels burning up.

Three days later, I'm out in the shop. One must throw off a flu quickly, before invalidism sets in. I don't feel well enough to chisel any huge notches in beam stock anytime soon, but a little delicate joinery always soothes my urge to get moving again. I want a nice slow tool in my hand, something that won't jazz my enervated body. No power tools.

My first project is a stopped dado, a slot in the bottom of a maple shelf to accommodate a bracket. Vertical chiseling, cutting at right angles to the surface of the workpiece and usually across the grain, requires a perpendicular cut straight down—one easy bang with the mallet—followed by a slanting cut to form a triangular groove. The bevel of the chisel is always held toward the waste side of the cut; put the flat edge of the blade right on the pencil line. Once the lines are grooved, I lay a paring chisel between them, bevel edge down, and push out wood chips. A router would do it easily, and much faster, but the noise would rip out my nervous system.

My two favorite chisels are kept very sharp; these are the tools I reach for to plane out the odd dado and mortise. One is the aforementioned Stanley, which looks like a socket chisel but actually has a full metal tang from the blade to the cap, as a magnet proves. In the process of sharpening it, highlights appear from some truly tiny print, and I find that this paring chisel does not date from the 1920s as I had always assumed, but from 1909. No wonder I have never been able to find any of its mates.

The other one is truly huge, a rugged firmer chisel with a 1-inch blade, British-made, part of a set of nine that I bought twenty years ago. The others are around here someplace, but this one winds up doing the bigger jobs that the Stanley can't handle. I used it when I built the barn, and at the end of that job, it was still sharp. Good steel.

As a concept, chisels bear some thinking about. The true art of carpentry is in the joinery. That's what makes things strong, from furniture to houses. Chisels are the key to basic joinery such as mortise-and-tenons, from the big ones on rail fences to tusk tenons for a trestle table. You can square up a round hole with a chisel, and the V-bladed corner chisel will do a perfect job. For variety and challenge, there's always the handmade dovetail, the hobby of serious woodworkers, for which one needs the sharpest edge, the finest chisels, and a steady hand on a light mallet.

In the blush of returning health, I'm starting to feel really good about life and grateful for all my keen tools, until I think about Bill's barbecued chisels. The best part of a community is that it takes care of its own. Maybe we should take up a collection. I've got too many tools anyway.

There's no tool like an old tool.

ANONYMOUS

BENCH VISE

My Only Vise

"Not so," says Joy. "You have a vise. I've seen it."

Yes, stipulated, true. I have a machinist's vise. In addition, I have many personal vices, from biting my fingernails to ribbons and compulsively recycling to acquiring tools I desperately need and even some I may never use, just for the sheer joy of putting them in working order. And speaking of Joy . . .

"You say this other vise would be for holding wood. And the vise you already have is for . . . ?"

I clear my throat for the first hurdle. "Lots of things. But it's a machinist's vise, they only *call* it a bench vise, it's not really a—"

"It's a big vise. What do you use for a wood vise now?"

There we go. "Nothing. That's what I'm trying to tell you. I need a wood vise. I *need* one." For choice, a new one, expensive and precious; my eyes have been peeled for a used vise, but they never turn up at sales.

"I mean, ask any woodworker if a woodworking vise is essential, and see what he says."

"Or she."

"Or she, yes. Good point. You could use it anytime."

"We can't afford it," Joy says.

"I can't afford *not* to have it. The Workmate can clamp small things together, and yes, the machinist's vise will hold a board, maybe, but nothing bigger. You can't count the pin vise because that's for teeny stuff. The saw vise will hold handsaws and nothing else; I broke the last one trying to make it grab a board. And the pipe vise is for pipes only. Or any round stock."

"Like dowels? They're made of wood. I mean, is this other vise really a priority?"

Priority: n., from the Latin *prioritas.* (1) Antecedence; precedence. (2) A first right based on need or emergency.

I've managed without a woodworker's vise all my life.

"Not . . . exactly."

How to put it? A woodworker's vise, even if I never used it, would add an air of legitimacy to my wood shop. People would enter this summer and behold this beautiful vise, and by damn, they'd know it was used for serious woodworking. The model I'm looking at has a built-in dog, quick-release jaws; it's made in England . . .

"Now, as I understand it, the vise you're looking at costs over a hundred dollars. How much over?"

Here we see two apparently opposing philosophies; but, on closer examination, the underlying similarity of viewpoint far outweighs the iota of difference. Joy and I both are willing spend extra for quality. "Ten dollars more. I, uh, seem to recall how much we paid for your Cuisinart."

Joy's eyebrows ascend. "Oh, that's different. We use that every single day, for chopping, slicing, dicing . . ."

"Yeah, but then, that thing they sell on TV does the same . . ."

My beloved gives me a look: *Nice try.* "There's no comparison. A Cuisinart does twice as much, ten times as fast, and better. It's a reliable machine. I used one in the restaurant." Joy once worked in one of the finest vegetarian restaurants in Vermont, and she knows I can deny her nothing in the way of kitchen supplies. Not if I want to eat well.

Well, I know when I'm beaten. Besides, she's right, it's not a priority considering all the other hemorrhaging holes in our June budget. "All right. I can wait another year for the vise, just get by on what I have. But honey, it would be nice to get one before summer is over. I could build new cabinets for the pantry." A vise is a sine qua non for cabinetry; tell any cabinetmaker that he or she must part with the old bench vise, and watch how he or she beats on your head with a lignum vitae mallet to let in some daylight. You'd have to be nuts, or poor, not to have a woodworker's vise in a woodworker's shop, like mine. Exactly like mine. Well, the pantry cabinets are a bargaining chip, a quid pro maybe-someday-if-I'm-lucky quo.

Meantime, I'll start on a bookcase, limping along another day without a vise. The first month of summer will be a challenge, considering how much cabinetry I've got to build, and when the time comes to rub a glue joint—you glue both edges of a board, put one board in a vise, and rub the edges together to squeeze out the excess glue and smear it well into the pores—I will need seven arms. But I'll do it, since I have to.

Here we have the stock, selected from the finest kiln-dried alder. The first item is to dress the edges of my two long-side pieces, between which the shelves will fit. If I had a wood vise, I could plane it standing up. Instead, I clamp them together and rest them on edge on the concrete floor of the shop, padded by a wide plank. Then I put on my knee pads, double damn it. Reluctantly, I straddle the doubled board, knees clamped tightly on either side. This is keen, just the way cave individuals did it during the dawn of time. Good thing my old boss can't see his first apprentice groveling on the masonry like some naked Third-World artisan. He'd say, "Buy a wood vise right now." I know he would.

Planework isn't the only thing I'd use a wood vise for. When using a drawknife to form a curve in a board, first you clamp the workpiece in a woodworker's vise. Ditto with a circular plane, which has a flexible steel sole. Once you have the wood held fast, you can do anything with it.

Maybe I'm thinking too small. What I need, if we're dreaming, is a full-sized cabinetmaker's bench with a pretty wood vise on the side. What I have is a starving writer's bench (vise not included), massive, serviceable, and dog-ugly. And a mechanic's vise, mounted on a stand.

But enough whining. My jack plane takes off long shavings of wood, although I have to remember not to pinch my finger or accidentally sever my femoral artery. This isn't overly safe. Wait a minute—yes, I'm sure we have plenty of gauze compresses and tourniquets in the first-aid kit mounted over there on the wall, two feet off the floor. Visitors ask me why it's so low; I have to explain that one reason is in case I need to crawl to it, unable to walk from loss of blood. People who seldom use tools don't always understand the need for extreme caution around sharp edges, but that medical kit is a good visual aid.

The word *vise* comes to us from the French *vis*, meaning *screw. D'accord*, I can grok the symmetry, down on my knees here. The British spelling is *vice*, which lends itself to all kinds of confusion and punnery. Some vises have jaws that come together by means of a cam or lever, and some of the earlier models had jaws, sliding mechanisms, and even screws made entirely of beechwood. Different trades made modifications on the basic tool: a coachmaker's vise, for instance, was mounted above the working surface of a freestanding pedestal or bench, with clearance all around it for trimming felloes, but a cabinetmaker needed a pair of vises, both flush with the bench top.

My dream vise has jaws (sometimes called cheeks) that can be faced with wood, and the whole machine set into my bench flush with the working surface. It has a dog, sort of a square metal peg, that can be raised to provide an opposing force against a bench dog or lowered out of the way for the vise to do its thing. The whole point of a vise is to hold and stabilize the workpiece at a comfortable and safe height.

In a pinch, I've made an on-site vise out of two pieces of 2 x 4 nailed to the deck, as far apart as the thickness of the workpiece—usually a two-inch board—but not quite parallel. I was able to slide the board into the tapered gap between the two vise boards, and held it in place with a wedge. This method works only slightly better than what I'm doing now.

I once met a blind woodworker. He wasn't completely in the dark—that is, he could make out shapes and shades of lighting—but he was legally blind, so the phrase "uniquely challenged" probably applies here. His workshop had been organized to a degree that I will never attain in this incarnation. Everything not only had a place but it was the most convenient place for it. A lot of thought had gone into the ergonomically correct hardrock maple bench, with drawers that pulled out noiselessly on waxed wooden slides and a tool cupboard with a spring-loaded door. I gave that one some thought, and understood why: if he forgot to close the door, he might trip over it. Had I the brains God gave a guide dog, I would have asked him for the plan of that workbench.

What shines out now in the foggy corridors of my recollection, though, is his vise. It was a Jorgenson or Record or Tucker, a good one. Someone else had fitted it into the bench to his explicit specifications. It was a good job; the hardwood-faced jaws were so well matched to the surface that the entire vise seemed to be a part of the bench itself. You could barely see the lines. So to speak.

We had a nice conversation about the rewards of working wood. As we talked, he kept on working. He had clamped a piece of walnut between the long cheeks, and to plane it, he ran his hand over the knobby surface, checking the radial marks left by the saw that sliced the plank from a tree. A few passes with the plane and he ran his hand over it again, fingertips locating every high spot and dip. Watching him, I learned to do the same. When smoothing wood, touch is more reliable than sight.

His hobby was making furniture that was remarkable for its tactile beauty and fine joinery, glassy smooth to the touch even in places the eye would never find, such as the underside of a cabinet. Without a vise, whose jaws he padded with sheepskin for sanding, that kind of workmanship would not have been possible.

One other unforgettable: An expensive table saw sat in the center of the workspace, a really good commercial machine with all the extras except a blade guard. Its multitoothed blade rose from the polished steel table like a shark fin; I closed my eyes and tried to imagine cutting a board that way. Not for any money.

Speaking of money, it seems odd to me that our budget is on the ropes to such an extent that I can't buy a lousy wood vise. Sure, the plastic is maxed out, and it's been a while since my last check came in for an article, but Joy's adamancy this morning seems out of character. A little begging and pleading usually brings her around, but not this time.

A week goes by, and another. Grumbling, I have to put off all sorts of projects for lack of a wood vise.

On Father's Day, I discover why Joy dug her heels in when I wanted to buy a vise. She and Ren awaken me with breakfast in bed, and they're both carrying a big box that weighs about twenty pounds, hot off the boat from England. Inside is a Record woodworking vise with 9-inch jaws, adjustable dog on the outside cheek, a clever quick-release, the whole body made of Sheffield steel.

"You sure are gullible," Joy says. "Did you really think we couldn't afford a vise? You can't afford not to have one. That's a direct quote, I think. Now, honey, about those pantry cabinets . . ."

It is easier to suppress the first desire than to satisfy all that follow it.

BENJAMIN FRANKLIN

AWL

George on My Mind

I've probably worked with five hundred people over the course of my life as a carpenter. Most of their names are lost to me, and their faces are blurry when I think back. But I clearly remember George's name, face, the way he talked, and one of his tools. Whenever I pick up an awl, I hear him humming.

George was a good carpenter, fast and professional, easy to work with, and it's strange that he had any enemies or that there would be people who wanted to deny him a job on sight. But of course there were. George was black, the color of coffee with a dash of cream. Some people had no problem with it, but some did.

As late as the mid-eighties, minorities were a rare sight in the building trades, which, for that matter, still aren't fully integrated. Now and then you'd see a face that wasn't snow-white, but considering that construction required no higher education and therefore had members of every other socioeconomic class, you'd think there would be more room for disadvantaged Americans.

But prejudice kept blacks and Hispanics completely excluded from the building trades for a long time, and back then it was common knowledge that women couldn't do the job, although in 1995 they seem to have no problem doing it. I remember reading about one of the first women to join the carpenter's union in Portland, in the early seventies; the first week, two Neanderthals broke her thumbs with a hammer to discourage her. (This species of ape is nearly extinct. Evolution takes time.)

Gays fared a little better, as long as they kept an utterly flat profile or could fight back. I saw an ugly fistfight once because one carpenter called another a queer; the accused carpenter mopped up the other guy, who had swung first. But in over two decades as a blue-collar worker, I can count the black carpenters I worked with on one finger.

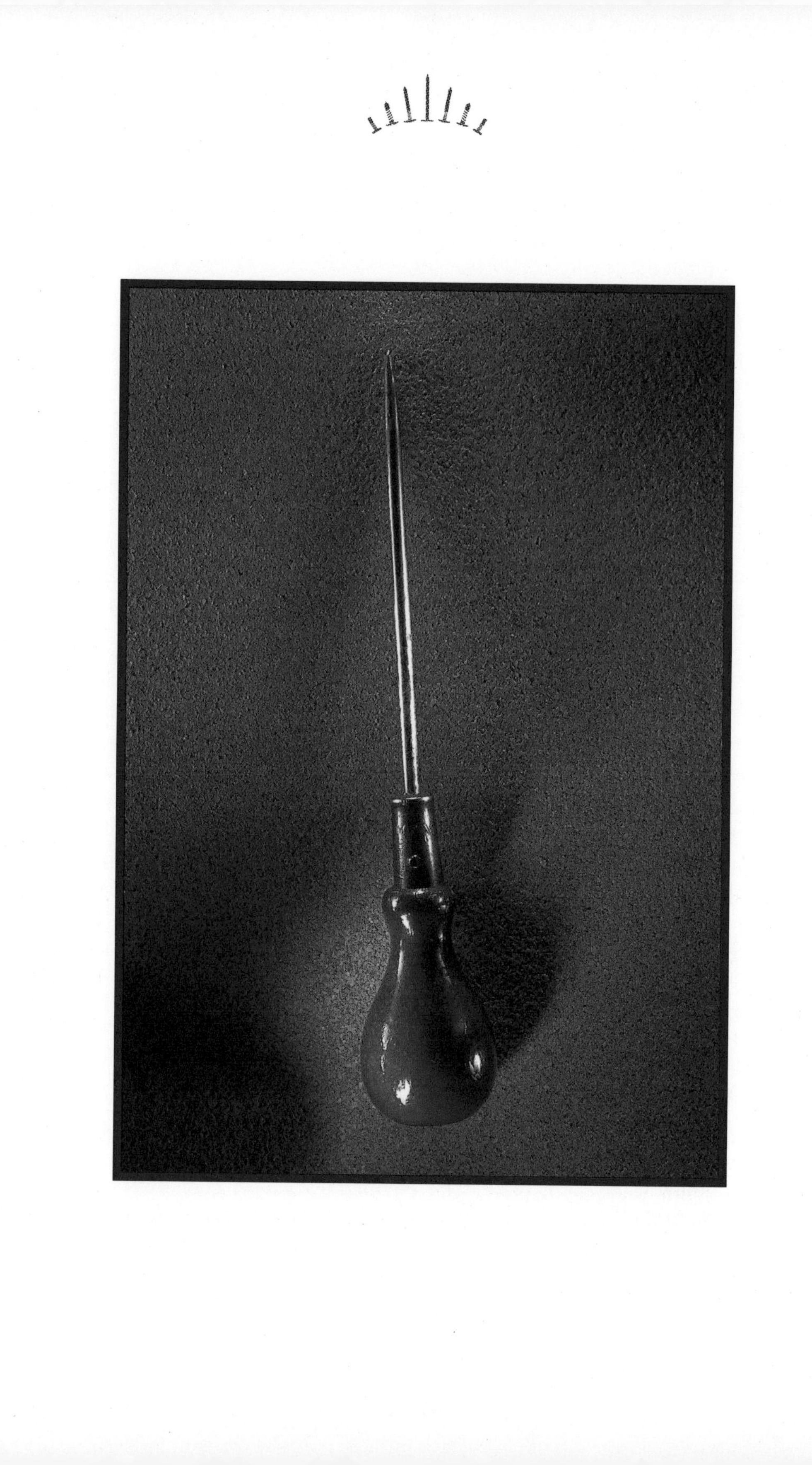

As everywhere, there was overt and covert bigotry. "You know where you stand with some people, right off," George said. "With others, it takes a little longer for their hate to wake up. Then you're 'lazy.' Overnight and all of a sudden," he added.

I met George on a finish crew in Colorado, one long-ago summer. We were "trimming out" an apartment building designed for student housing near a university. The specs called for colonial molding, the pretty stuff with flutes and ogees, a step above the usual flat sanitary trim. The interior doors were hollow-core but paneled, an expensive way to go for apartments. The cabinets had been contracted by the finest shop in the city. All this meant that the general contractor wanted good carpenters who could cut perfect miters and hang a door so the reveal would be perfectly even all around the edges.

The finish crew was large, maybe twenty individuals, and we were divided into two-person teams—two-man, really, since there were no women working on the site, unless you count the secretary in the job shack who disbursed our paychecks every Friday. I was teamed with George, who must have been about thirty-six then, ten years older than me. He was a master carpenter who had worked all over the West, especially during the boom years in California. His father and grandfather had been carpenters, back in the days when "the only work they could get was doin' for other people of color," he said. His voice had a remarkable clarity and timbre, a rich baritone; he sang in his church choir, he mentioned once. He had spent a lot of time in Texas and the Great Lakes region, so his speech was an odd drawl that melded suburban Houston and downtown Chicago. He spoke loudly, booming out words from his diaphragm.

"I never talk soft," he told me once, his big mustache rising as he grinned. "Not no more. Y'all gonna hear me coming."

That first day, we put up cabinets together, following the blueprints. When you work with someone, watch the rhythms. Both of you should work at the same speed, and if both of you know what needs to be done, communication can seem like a form of telepathy. Without discussion, I knew when I was supposed to be holding a cabinet or taking a measurement; and when I was ready to drill holes for the screws, George had already marked the stud placement. He hummed while he worked, sometimes singing a few bars of a deep and lonely melody, or adding music to a sentence. I can still hear him singing "Jesus, won't you tell me where I put my hammer" when he couldn't find it.

By the end of the day, we had the rhythms down. "You done this before," George said as we picked up our tools. It was a subtle compliment.

The next day, George began hanging doors while I trimmed the windows. I noticed that he used his awl to make pilot holes for the lockset. It's not easy to do without splitting the wood.

To install a lockset on an interior door, you need to bore two pilot holes for the latch-bolt and two more for the striker plate. There were five doors to be hung in each unit: one entry, one bath, two bedroom, and one closet. Multiply five doors by the number of apartment units—if memory serves, there were around two hundred—and that's how many pilot holes would be needed to put doorknobs on the doors. Most finish carpenters carried some quick device for drilling or punching the pilot holes. Some carried little gimlets, others used a center punch, as I did, and carpenters like George simply placed the point of their awls in the correct spot and banged it lightly with the heel of their free hand. Presto, instant pilot hole.

Every so often, he pressed the point of the awl into a block of beeswax. I noticed that after he struck the awl, he removed it with a little twisting motion, polishing the sides of the pilot hole. He used a Yankee screwdriver at full extension to set the screws, which he also dipped in wax. The screws bored in easily, without splitting the jambs.

"My old boss swore by that stuff," I told him, pointing to the little wax ball. We were both working in the living room, ten feet apart.

He smiled. "Man sound like a genius." He made the G very soft and prolonged.

"He was." We worked for a while in silence, and then I asked, "Hey, how'd you get into carpentry?"

He laughed, a rich yuk-yuk that sounded like it came from inside a cave. "They always ask," he told the door jamb. He placed his awl in the striker mortise and bopped it twice with the other hand. "What y'all mean is, how'd a nigger get this far?"

To me, the word sounded like a pistol shot in that room, although he put no special inflection on the word. But apart from that deadly word, I guess that's what I meant. He was an anomaly. I didn't know what to say, so I said nothing.

George kept working, humming softly. After two or three minutes, he spoke. "It's like this. My daddy said I should just show up and keep showin' up, and someday some fool would hire me. That's what I did, and that's what happened."

We didn't say much to each other until lunchtime. Then he lit a cigarette and told me a little more of his life.

He had learned everything he knew from his father and grandfather. It hadn't been easy for him to get and keep a job as a carpenter. He'd met prejudice and provocation, and when he did get a job, he'd been laid off when no one else was. "But you can always find work, if you'll take it. I've been a laborer when I couldn't get a real job." Laborers were the very lowest rank of blue-collar, the unskilled workers who ran shovels, pushed wheelbarrows, carried steel reinforcing bars for concrete, and swamped out construction debris for low pay.

"Had one job in Texas where they tried to scare me off," he said. "Two big fools come over while I'm doing my job, a big one and a mouthy little one. They eyeball me for a while, but I just keep on working. Little shrimp says to me, 'Nigger, you just quit.'" George pulled his awl from its scabbard on his toolbelt and held it out at arm's length. "So I whipped out my little cracker-sticker and put it right under his eye. 'I'm staying,' I says. And they backed off."

Until that time, I had never thought of an awl as a defensive weapon. It's a handy tool for making small holes in wood or leather (superb for adding another notch to the toolbelt), and it's often used to align holes. The handle is round for an easy grip, with a flattened side so it won't roll away when you put it down. The blade can be anywhere from three to nine inches, and the point is either a standard needle or four-sided.

A scratch awl probably preceded the original drawing knife as the original tool for scribing a deep line in a plank for a handsaw to follow. This tool's blade is set into the handle, usually beech, and later types such as the punch awl featured a full tang that provided a metal knob at the top, a place for the mallet to strike. For the most part, though, an awl is a hand-driven tool. It slides into its scabbard on the toolbelt and occasionally stabs you in the leg when you crouch.

George and I worked together only three more weeks. One day he was gone. I asked at the office and found out he'd gone back to Chicago because his father was dying. They paired me with another man and work went on as usual, although the new guy insisted on setting up a radio so we could endure hardcore country western music all day long. After the first hour, I put foam plugs in my ear and pretended George was humming.

On payday, I went to the job shack to get my check. I asked about George. The superintendent shook his head and chuckled. "No, he isn't coming back. He called and said he found another job. Probably got himself arrested by now." The super collected some papers and stacked them up. "That's all right, he was getting lazy."

Just for a second, I wondered if an awl would be long enough to reach his walnut brain, if you stuck it in one ear. Probably not.

I never saw George again. I hope his life in Chicago turned out okay. No doubt it did. He said something to me during our brief association that seemed to sum up his philosophy: "You never know where you're gonna be tomorrow, or who you're gonna be with. The wind bloweth where it listeth. Every day, you gotta take a big drink of life, and drain the honey from it."

A tool is but the extension of a man's hand.

HENRY WARD BEECHER

KEEL AND PENCIL

Making Your Mark

I must have been ten or eleven years old when Sergeant Friday was falsely accused of manslaughter and asked to hand in his badge and gun. I witnessed the whole thing on black-and-white TV: A criminal had drawn and fired in a small hotel room; Joe had returned fire in self-defense, blowing the perpetrator away. But the investigators couldn't find the bullet he claimed had whizzed by his square head. Ergo, not self-defense.

It looked bad for Joe Friday. A worry wrinkle crawled briefly over his brow, a real wave of emotion for Joe; up the river, he'd have to hold the soap bar in his teeth, just to be safe. But then the cops gave the room another fine-combing; this time they found where the bullet had gone under a shelf, lifting it slightly off the cleat. When the shelf dropped back in place, it covered the bullet hole, leaving only a tiny line on the bottom to mark the round's passage. "They thought it was just a carpenter's pencil mark," his partner (Bill Smith—pre–Frank Gannon) explained. "One for the books, Joe." Sergeant Friday shrugged, and they gave him his badge back.

For me, this episode answered an important question: How carpenters knew where to cut. They made a mark with a pencil. I had lots of pencils in school, but I wasn't a carpenter; and no one called them carpenter's pencils, so that item must be something special, I thought.

The next time my father took me to the lumber store, I asked the man behind the counter to show me a carpenter's pencil. It was yellow, like mine, about the same length, but flat. I asked why. "So it won't roll away," the man said, "when you put it down."

It all made sense now. It was my first prickle of interest in the field of carpentry. A few years later, a high-school shop teacher scotched it, thereby throwing my life direction into question for a while.

Good old Mr. Grutlusker. He looked like Jack Webb, a little, although Webb had a kinder face. The first two weeks of shop class, he made us practice drafting with common pencils, exploring the topography and measurements of a melon-sized dodecahedron with a hole drilled into the center of one face. The drafting paper was very soft, and the pencil lead was very soft, and the rulers had nicks in the edges. The entire exercise was a refinement of torture. Smudges and erasures and blips on a pencil line counted against us. By the end of that time, I was sick of pencils, objects with dimensions, and Mr. Grutlusker. I forget what shop project came next, but it was five years before I wanted to get intimate with another lead-filled pencil.

Swanee fixed all that. Maybe you wouldn't think a carpenter's pencil mark could be made with artistry, but Swanee did it. In the hands of a real craftsperson, there's a sound a pencil makes when it draws a line along a square. Amateurs strike the line two or three times, *zip, zip, zip*, but a professional draws one line: a controlled *sssst.* Swanee's yellow pencil was an extension of his finger; when he made a mark on a board, he barely touched the wood with the graphite, and yet it left a clear arrowhead, a point to be bisected neatly by his square. His pencil poised to make another mark, this time a line, drawn with no hesitations or flourishes, distinct but not deep enough to mar the wood. Even on drywall, his pencil marks were light but clear. No one I ever met afterward made marks that expertly, focus without apparent concentration. It was like watching a Japanese master calligrapher write the ideogram for "bird."

Swanee used only one kind of pencil, given away by a lumber store clear across town. (Like road maps, at one time a freebie at every gas station, a good carpenter's pencil used to be given away by every hardware and lumber store. That is no longer the case.) I never found out what made his pencils write so well, or who made them. At the time, it didn't seem important.

A carpenter's new toolbelt has pencil pouches on the left and right. After a while, the pencil slides in and out easily. If the hammer loop isn't in the right place, or if the pencil is in a slot that should hold a nailset, the hammer head can snap off the point of the pencil; until the pouch is properly broken in, you move the pencil around a bit until it finds a home it likes. After a year or so, it won't fit in any of the other pouches.

The science of making stud marks on framing lumber is called "layout." When laying out a wall, you stretch the measuring tape down the doubled plates—they're tacked together on edge with small nails, so you can lay them both out at once—and make a definite V-shaped mark, a caret, every sixteen or twenty-four inches. Some carpenters make the stud line right then, with a square; others go back and do it after all the layout marks are made. (Choose method #2. It's easier.) Then you make an X on the same side of every line, or a T to mark the trimmers (the shortened studs that support window and door headers), or a C to mark cripples (short studs under and over openings in the stud wall, for windows and doors).

That's how it's done. The same basic method is used for marking off the cap plate and ridge pole for rafters: make a little mark, draw a square line at that point, and make a big mark to indicate which side of the line the rafter goes on.

Sometimes an inexperienced crew follows the layout person; sometimes a placement mark is barely visible under the glare of the sun, or goes dim from too much fatigue. A big penciled X can be harder to see than you'd think. Problems ensue: Studs can be omitted, put on the wrong side of the line, or placed where cripples should go. A pencil mark may stand out on the fine grain of interior wood, but on planed two-inch stock or plywood, only the softest lead will leave a good fat mark. Soft lead won't groove the wood; for that very reason, cabinetmakers and trim carpenters prefer the softer, erasable leads. A hard lead doesn't need sharpening every five boards, so the framing foreperson will use the hardest pencil he or she can find. Eyestrain is an occupational hazard of framing.

That reminds me of Keith, who ran a crew of framers without saying a word. Some of his carpenters were experienced professionals, others were summer amateurs. When I worked for him, he had solved the eyestrain problem by making the X (or T, or C) with a blue carpenter's crayon, sometimes called a keel. The keel fits in a special holder, a tube made of end-slotted wood with a metal ferrule that tightened it down to hold the crayon. I once asked Keith: Which part was the keel?

He had the same answer for everything. Keith was afflicted with a stutter you never noticed unless he was talking to you, a problem he obviated by avoiding speech, including explanations of any kind. His marking technique made things simpler, with no discourse or commentary required. He always did the layout in easy-to-see blue crayon; there it was, big as hell, and what possibly was there to talk about? Ask him a question and he'd shrug and look away: *Don't ask me.* I liked him a lot.

I noticed that his method of layout marking avoided errors, so when I had to lay out for an inexperienced crew under a hot sun, or under any other circumstances, I also used a keel to make the X's. But wait—

Technically, "keel" refers to the crayon itself, not the holder, which is cleverly called a "lumber crayon holder." But here are two problems: (1) the average hardware store clerk has no idea what a keel is, and (2) what do you call the device in your hand when the crayon is inserted, protruding one inch out of the end? A keel inside a holder? No. You can call it a keel, the whole thing, and no one will refute you. Take it a step further and call the naked crayon a crayon (a French word that means "pencil"). So if a bare keel is a crayon, let's call it that, leaving the name "keel" to refer to the whole assembly.

I still use a keel, even when I'm the only one reading my own layout. This is partly due to my eyesight, which is not what it used to be, but mostly because I found the most beautiful keel in an antique shop. It occupies a place of honor in my toolbox: right on the lid where I can always find it.

Go into a hardware store and take a close, critical look at the model they'll sell you: The wood is mahogany, the ferrule made of thin steel. A plastic cap at the end seals it off. It's okay, I guess. The same company who makes the crayons makes the holders.

Now examine the one in the photograph. The body is made of ash or hickory, stained a dark rosewood hue by decades of use, and that's real brass at the cap and ferrule, and a leather loop. The "Peterson Crayon Holder" was made by "T. A. Peterson, Longview, Wash. Patented." Now there's a keel. He was proud to put his name on it. I'm proud to use it.

You'd have to haunt a lot of antique stores to find one because carpenters tend to bequeath their finest tools to relatives, who hold onto them for years for sentiment's sake. Maybe it's best just to forget the older tools and buy new ones of the best quality possible. The problem is, while you can find new presentation-quality awls and marking gauges in the fifty-dollar range, both with rosewood bodies and brass fittings, no manufacturer will ever again produce a keel quite this beautiful, handmade by craftsmen—not in this century, anyway. There's no call for them. Money can't buy them.

On the other hand, I found a second one the other day, nearly identical to my old favorite, after only ten years of looking. So there's hope; you just have to keep looking. I'll give it to our daughter in another twenty years or so. It's an heirloom.

There is no new thing under the sun.

ECCLESIASTES I.9

SPIRIT LEVEL

Centering

The process of finding out who you are, what you are doing here, how long it will take, and what you don't know uses up an entire lifetime. If you want to plumb the depths of your own ignorance, switching careers in midlife is arguably a smart move, especially if the original career puts personal survival in doubt. After a WANDE (Work-Associated Near-Death Experience) or two, you may realize that you've got to do something else for a living.

Usually it begins with an epiphany. I, for instance, was standing on a high ladder over a concrete slab in a large commercial construction site, watching carpenters running around like squirrels twenty feet below and ten feet above, when the concept struck me: Someday I should write. This decision was driven home by a falling ten-pound drill that slammed into the side of my skull, inflicting great eyelid astronomy.

Luckily I was wearing a hard hat, and I stayed conscious long enough to slide down to earth, bleeding and smiling. My time wasn't up yet. It only hurt so much because I was alive. Best of all, I had just received a clear career insight: GET OUT.

That same day, the mail brought a check for an article I had written equivalent to my weekly pay as a carpenter, and what started as an avocation became my livelihood.

As I said, it was an epiphanic moment. Another was yet to come.

Most of my old hand tools have one thing in common: they were originally owned by people who were alive in the past but who are no longer breathing. Tools can't die. They just keep moving from hand to hand, down the years. It's something to ponder, that all the well-made things we own will outlive our bodies. Maybe we will, too.

It was autumn in the mountains of New Mexico. Five of us were building a house, a chalet-style two-story with a full basement, very plush but a mere second home

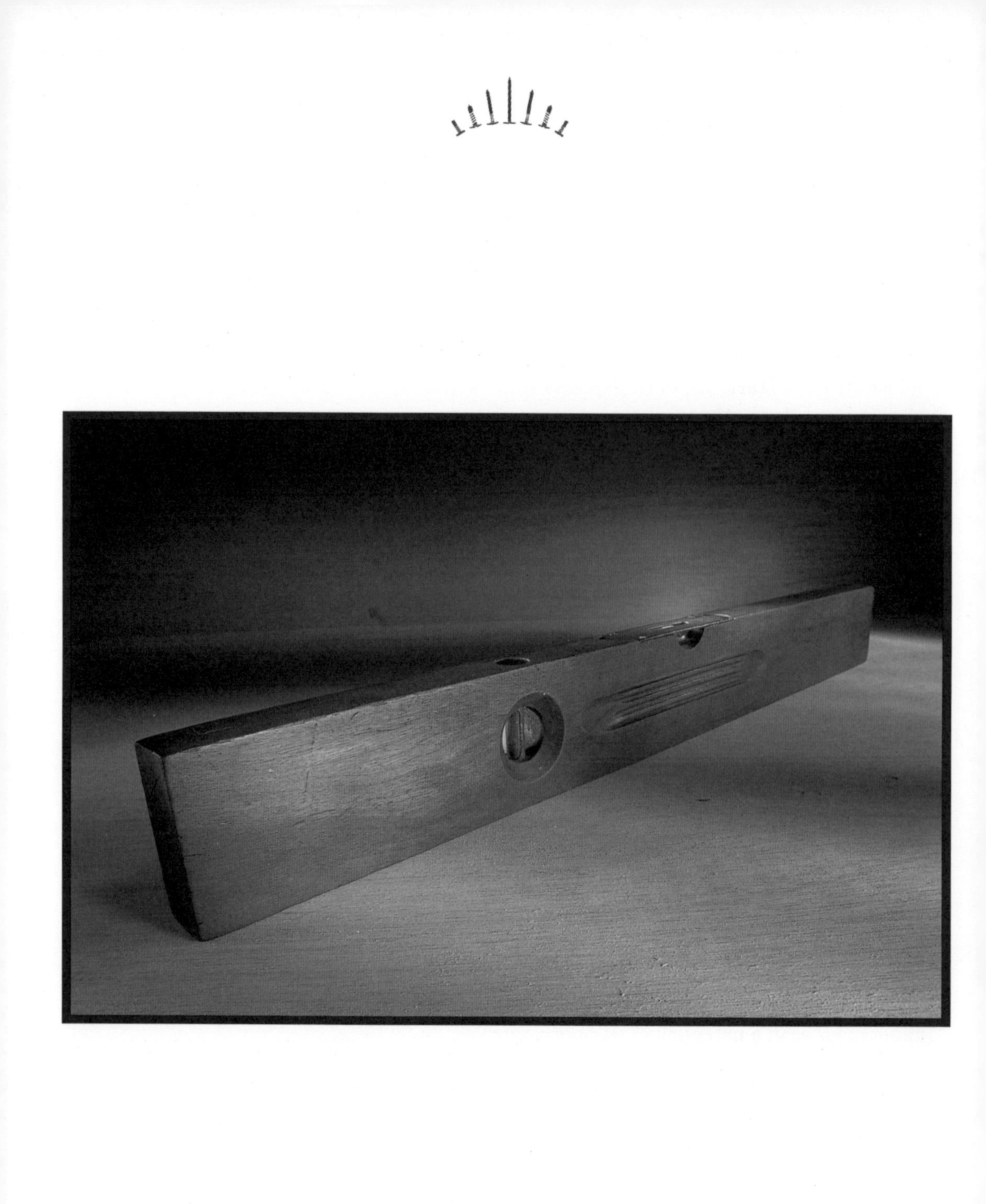

for the owners. They had bags of money, a personal jet, and various itches that money wouldn't scratch. Mrs. Owner refused to reconcile her continually evolving dream house with the stark fact of construction, and kept asking for plan changes well into the framing phase. Mr. Owner was an apparently religious man who drank himself into oblivion at least once a week, speaking fondly of the pending afterlife in Heaven until he lapsed into a boozy glossolalia. Their oldest kid, Bud Owner, Jr., was serving a few years for drug possession and other crimes against his mind. "The ungrateful little shit," Mr. Owner said of his son. "Jail will do him lots of good." Even with all their riches, they were a pitiable clan.

By contrast, their carpenters enjoyed an idyllic existence in a supportive environment. Our crew had been together for over a year. We traded stories and histories, invited one another over for barbecues, listened to the same music all day, told jokes, and taught one another the little tricks of building. We worked outdoors on the sides of high mountains, with a view to die for.

Given enough time and good chemistry, a tight carpentry crew becomes a loose family. Mike Hammer, the graying foreman, hailed from the East Coast but had built in the area for years. His partner, Gary, was also his younger brother, the ramrod whom Mike trusted to lead the crew in his absence. Last-to-Come Lee, a stout Native American, had the least building experience but fantastic stamina and a real give-a-damn sense of humor. Jack Wayne was a recent college graduate, even more recently married, and so desperate for a living income that he stayed on at his summer job. He was prematurely bald, wore thick glasses, and looked like a philosophy professor. And there was me, the from-Iowa, Oregon-bound carpenter who jotted down random ideas in a notebook at odd moments.

Mike and Gary set the tone for our crew. They were professional contractors who wanted to do a good job, safely and on time. Jack and I wanted to know every detail of the process because we both hoped to be contractors ourselves someday, and we pumped them ruthlessly for information.

Lee, older than any of us but Mike, kept us all from getting too serious. Last-to-Come was Lee's real name, or at least the one he gave out to the public. ("Don't worry, white man, you'll never know my real name," he told me.) And in fact, he was the last to join our crew. He didn't have many tools: a hammer and a toolbelt, a battered worm-drive saw, a leather bag of pliers and screwdrivers, and an old spirit level, a real antique.

"It's a sacred instrument," Lee avowed one day after work, running his long fingers down the grooves in its side. "First thing I bought when I landed back in the World." Lee was a Vietnam vet, brimming with stories of the A Shau Valley, 1969, and he intended to write them down someday. "This here stick shows how water flows and the wind blows. Look at this little bubble, right through here. Isn't it cute? I'll tell you what it tells me: the balance of all things."

You have to trust a level or you may as well not use it. A spirit level gets dropped, used as a straightedge, tossed in the back of pickup trucks, and generally ignored and abused until it's time to use one. Then the carpenter fully expects the little bubble to report if a wall is plumb with the center of the earth or a sill is level with respect to the even forces of gravity.

Electronic levels seem to be taking over the world these days, and to tell the truth, they are ultra-accurate. Digital displays tell the user when level or plumb is exactly on, zero degrees perfect. The brains are in a little "module," which can be attached to any length of bar. Replace its batteries now and then, and you have a level made for the Information Age, a tool that gives you raw data.

Even so, these levels are dead to the touch. Every part is artificial. No dryad ever lived in the cherry tree they were not made from. No sunlight ever found its way down to their cells. Yours will be the first hand ever to inform them, and when you die, they will still be ticking off numbers that any idiot, or even another machine, can read.

Spirit levels got their name from the fluid in the vials, usually alcohol because it doesn't freeze. All of them work on the same principle: when the bubble is centered between two lines, the body of the tool will be level, or plumb. Carpenters use six- or eight-foot levels to plumb doors and house corners, and shorter levels for everything else. Most are steel or aluminum, although a few old wood versions are still floating around. A mason's four-foot level, for instance, is made of mahogany, brassbound at the edges and corners. But not for much longer, I'll bet. One word: plastic.

The day Lee died started like any other. It must have been a Wednesday, the September sky clear with a few fat cumulus clouds, the air cool, dry, and thin. That day we were cutting rafters for the complicated multigable roof and setting them in place. We'd been in a happy mood the week before, laughing and taking it easy, but now we moved with a purpose. Gary had told us that Mike was sleeping badly, worrying about all the plan changes and the cold weather coming in a few more weeks. We had to get this job done.

But it wasn't hurry or carelessness that cost us a friend. It was bad luck.

When it happened, Mike was in town, meeting with the owners. Gary kept the pace up, perhaps trying to impress his brother with how much we could get done in one morning. So were we. Lee had become positively manic about it, humping rafters up to Jack and I as fast as Gary could cut them.

"Come on, put 'em down, Jackson Browne," Lee said. "You should get a bigger hammer, like mine."

"Hey, you can eat mine," Jack suggested, pretending to throw his hammer. "You savage."

"Savage? Oh you bitch, John Wayne." Lee blew him a kiss. "Never fear, we're gonna wear you out so your poor wife gets a little rest tonight."

Jack gave him the finger and smiled, hauling up another rafter. "Last-to-Come," he said. "How apt."

By noon, Mike was still in town. We'd asked him to bring back some sodas, figuring he'd be back by the time we broke for lunch. "I'll go," Lee said. "I'm out of cigarettes anyway." There was a store down the road a few miles.

"Bring me back a pack of Camels and a root beer," Jack said.

Lee brightened. "Hell, I'll bring you back a fresh scalp. We'll staple it to your head for you. You'll look just like Tom Jones." And then he drove away.

Mike arrived before lunch was over, but Lee was still gone. By one o'clock, he hadn't returned. Two hours later, Mike began to worry. "Guess I'll drive down to the store and see what's keeping him. Probably a flat."

Mike was gone a long time. When he returned, his face was chalky. "Lee's dead. Auto wreck." He breathed out a long sigh. "Fuck it, let's all go home."

Lee had been coming back from the store when an oncoming car straightened out a curve on the wrong side of the line. That driver was in critical condition, but Lee had been killed instantly.

We were off for four days. Lee had no relatives that Mike could find, though he tried. We all went to the Sunday memorial service; Mike paid for all the burial expenses. Monday morning we went back to work, like zombies. We finished the house on schedule, but it was a joyless job. At breaks and lunch, we talked about Lee. "That beautiful bastard," Jack said. He poured a little root beer on the ground, lit up a Camel, and silently offered its smoke to the four winds. None of us laughed.

On the last day, Mike gathered up Lee's small pile of things from his pickup. "I think Lee would want you guys to have his tools. If any family shows up later, you can give them back." Jack received the hammer Lee had said he needed, Gary got the saw, and Mike handed me the spirit level. "Goddammit, *take* it," he insisted. And I did. It was the first time I'd ever touched it.

It's the only tool I've ever acquired that gave me no happiness in the getting. It's a good old spirit level, a two-foot device for finding the level and the plumb of things, patented October 29, 1912. A tiny trademark symbol on the top brass plate identifies the manufacturer as Zenith, Marshall-Wells. It has two vials. Its body was shaped from a solid plank of cherrywood, signed and dated by its maker: "118-J."

A dozen two-foot levels have since passed through my hands over the years. But I've kept this one, as well as the memories of its provenance. It's not as light as aluminum nor as accurate as an electronic level, but it serves to remind me of matters more important than building level floors and plumb walls.

In the fall following the summer of my father's death, I was using this level for the thousandth time when I had a numinous experience. The time and setting were utterly prosaic, a late afternoon on the back deck of my house. It began when I felt a presence, as if an unseen hand had also touched the level. I had placed the level on a balustrade, and—this will sound dumb—it seemed to slide a bit too easily, gliding along the rail like a Ouija-board planchette. My hand came off it for a second, just to see if it would move on its own power. It didn't.

I felt moved to ask a question in my head. Don't ask me why. The dead were on my mind, that's all.

Why life and death? What's it about? And I waited.

Slowly, my daily mind drained away, replaced by a sense of expansion. There is something ineffably poignant about autumn, an ache that also comforts. It felt something like that: not sadness over the past or future, but a voiceless longing for the present. Not the fear that winter's darkness was falling, nor the hope of spring to come; but the beauty of one moment suspended in the twilight between seasons. It held me. My chest hurt. The mood was passing strange, yet there was something so familiar about it, so soothing. I never wanted that feeling to end. There were pangs, but they were pleasant ones. The colors of the day deepened. I saw a swallow flying toward the ocean, and its sweeping flight contained all the wings of the air. I saw what the Sioux meant by "seeing in a sacred manner"—the slant of sunlight held a tawny joy, the shadows a deep and melancholy sweetness. Everything in the universe was in complement: light and dark, earth and sky, living and dying, past and future, plumb and level. Everything was clear, holy, and in perfect balance.

And then it left me, as gently as a door closing, and I felt like I always feel, which is fairly boring and ordinary. I had no idea what it meant or whence it came. But for a moment, I saw the heart of things, a glimpse of worlds behind this one, less tangible but more real. All I understood was that life is too short and tenuous to worry about, and far richer with meaning than we can imagine.

Existence is beyond the power of words to describe.

LAO-TSU

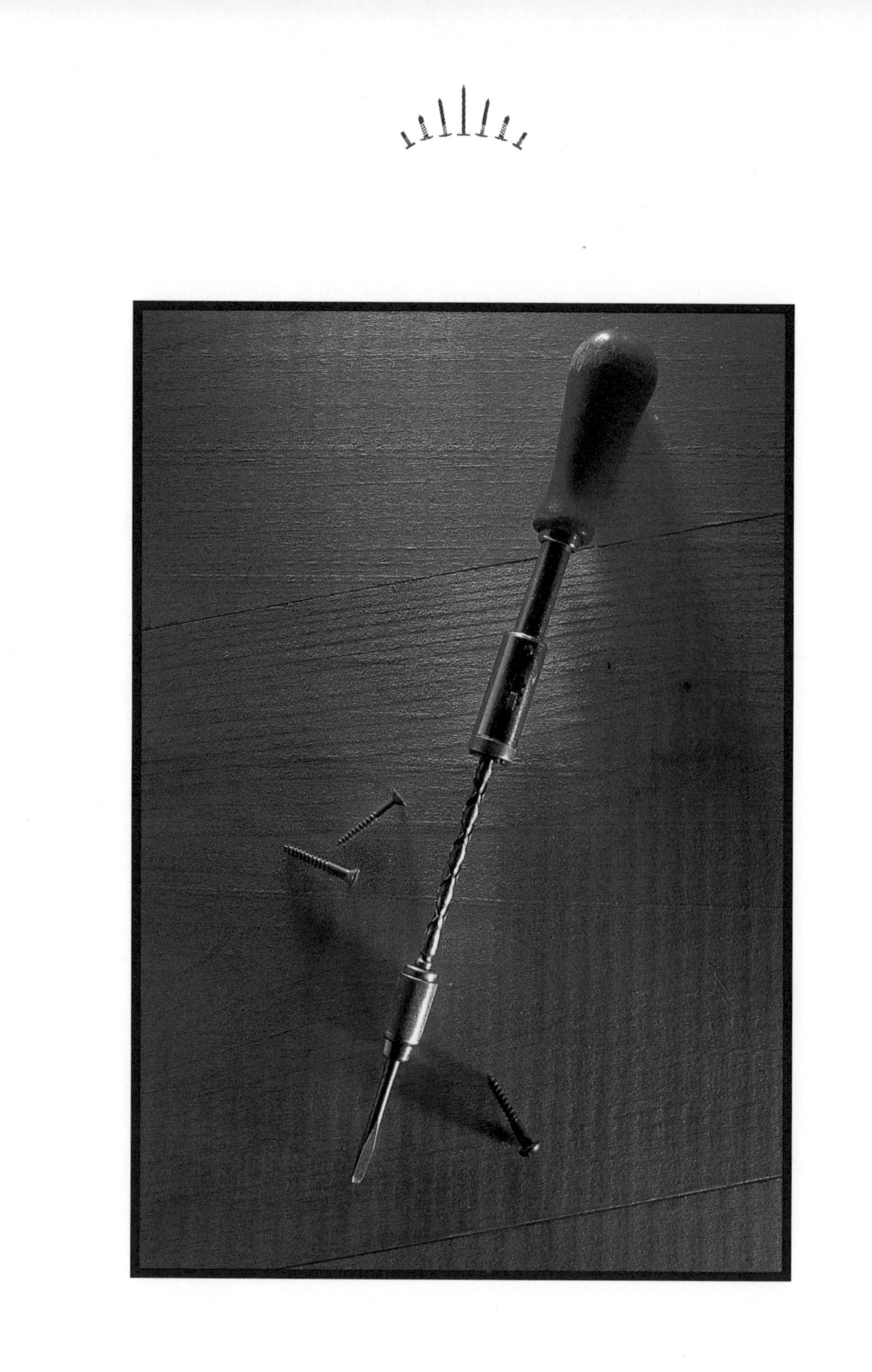

YANKEE SCREWDRIVER

Farewell to the Old

The older I get, the more I understand a few universal truths. Here's one: Change is inevitable. And another: Every gain carries with it a corresponding loss. But we're a very gain-oriented society, so in most cases we're happy to see the old pass away and the new take its place.

For instance, I just purchased the screwdriver/drill of my dreams, a cordless delight. In doing so, I lost the use of all my old reliable Yankees.

The progression is always the same with a new tool. The first thirty seconds of ownership are a rush, the sweetest, purest moments of nirvana, when a pristine tool comes off the shelf and into my hands. Then it comes out of the box, another happy moment but shorter and less poignant than before. I usually find a job for a new tool by the second day. After that, reality sets in.

Somewhere on the old homestead is a job I've been putting off until the right tool comes along, and it is no different this time. Down by the barn, an old gate sags on rusty hinges; I've been waiting for a way to drill some new holes for some new hardware. It's too far to run a power cord, and while my usual complement of hand-cranked cordless drills would do the job, I've been hoping for a battery-operated drill to come my way. Last week, one did.

From one perspective, my new cordless drill is just another tool, not unlike all my hand-powered drills. They all drill holes or drive screws. But this one is a 14.4-volt wonder; no power cords to string out, a big improvement in performance over most hand drills, plenty of torque, and it does the same job of work better and faster than all my others.

Here's the part that stings, though: From this time forward, the other drills and drivers won't get used much, if at all. My ⅜-inch drill has been sitting in its wooden box for over a month now, the cord wrapped around its body, and I don't foresee using it anytime soon. There have always been a number of cordless screwdrivers in my shop; now I have one more, better than all of them. It's a gain, and a loss.

Another thing: Half of my heavy-duty extension cords just became redundant. I won't need that many anymore. As further cordless tools replace those with power cords, they'll spend more time rolled up and waiting. I may even sell them.

The tools I'll miss most are my Yankees. Not that I'll part with any; I'll just miss using them.

I met the Yankee spiral-ratchet screwdriver in 1974, when I walked into a hardware store and bought my first one. There was a purple box kept under lock and key behind the counter, purple being the trademark color of the Yankee; in the box were dozens of ancillary bits. I purchased eight of them, including a countersink and a tiny reducing bit that slid in the collet of my new Yankee and allowed me to use tiny fluted drills. Then I took this tool to a new job, where it performed as promised, and after that did my damnedest to wear it out. It acquired a few scratches in the process, but it works as well today as it did when it was new.

For the next twenty years, I used that screwdriver whenever I needed a cordless screwdriver for drilling holes or driving screws. Over the years, several others of varying sizes and ages joined it in my toolbox.

No more. That's all behind me now. And I don't know whether to be happy or sad about it.

One thing's for sure. I've been denying myself a cordless drill for years in the interest of desire control. Now that I have one, I'll have to find something else, some expensive and useful tool worth resisting until I can't bear it any longer. It's a good exercise, almost a meditation. And just as that crazy farmer said it would, a little forbearance has always helped me to define "enough." The German word for it is *Spannungsbogen:* the self-imposed delay between the onset of desire and its eventual fulfillment. What a concept.

My newest cordless drill/screwdriver is a DeWalt. No one can accuse me of rushing into the acquisition of a new power tool. One comes to my attention, and I think about it for a year or so, biding my time and waiting for improvements to occur or for other manufacturers to produce something even better than the first one that captured my fancy.

That's what happened with this one, except that I waited five years. The first one that caught my eye in the store window was simple, powerful, clean-lined, and made by the same company that made several of my power drills with cords. I used a friend's on a small job, and it made quite an impression. I wanted one for my own, but there was no hurry, because I still had my Yankees. And I waited.

Now I'm glad I did. My brother Tim bought one of these, and my eyes fell upon it with an orgasm of covetousness. Even then, it took another six months before I got mine, a more powerful model whose battery pack also fits in a cordless saw made by the same company. By God, I bought one of those, too. Any company that makes a battery capable of powering both a handy drill and a lightweight saw is a company that has researched people like me. They probably interview two hundred of us and then go back to the drawing board. "If we build it," they say, "he'll buy one, eventually." And I did.

I have two Skil worm-drive saws that have been with me forever. They made most of the cuts for the house we live in, and scores of houses before that. I used them almost daily, wanting nothing better, until I got my cordless saw. Now they sit on a shelf, cords encircling them, in perfect condition. Every time I think about taking them down, it occurs to me how heavy and inconvenient they are. Then I pull the battery pack out of its charger and slap it in the end of my cordless saw, with hardly a pang.

There's no question in my mind that I may have used my Yankees for the last time. There are two sizes of Yankee on the lid of my toolbox, and now they're just decorative, or for emergencies. But sloppy sentimentalist that I am, I probably won't part with any of my five or six Yankees.

It must have been the purple handle that first stirred up hot winds in my tool-loving breast. I have an old Yankee with a wooden handle, all the paint long gone or never there in the first place, and it just isn't the same. In fact, the plastic handle on the newer Yankees (those made in the last twenty years) are stronger and feel better than the old ones made of plain wood.

The ratchet screwdriver was invented at the end of the last century. Two of mine were made by the North Brothers Manufacturing Company of Philadephia, which later became a division of Stanley Tools; a patent history is engraved into the body, and the most recent date is 1923. These tools have over seventy years of history behind them, of which I can account for only the last fifteen.

They still do what they were made to do: insert or remove screws, according to the position of a little rectangular button on the body. A twist ring at the base of the spiral ratchet locks the bits down, spring-loaded and ready to zip out. When we were first married, Joy was holding my biggest Yankee when she turned the collar accidentally. The sharp screwdriver tip missed her face, mostly because it was busy crashing through a brand-new double-pane window.

We started a small contracting business by the sea, and Joy worked with me until her pregnancy began. Yankees in hand, we remodeled houses and built stairways down to the beach, constructed outdoor furniture and kept funky old seaside cottages from falling apart. Most of that work has been bulldozed away, replaced by modern condominiums and restaurants with ocean views. I lost a Yankee on one site, probably buried behind drywall or fallen under some cabinets, and consoled myself that someday another carpenter would get a pleasant surprise. But the dozers came and tore it all down, loading the debris into dump trucks. No one will ever find that Yankee.

About a month ago, we put our house up for sale. We've lived here for years. We had our wedding reception in this kitchen, and our daughter was born under this rooftree. My father's ashes were scattered in the little river that runs behind the pasture, the Tum-Tum, which means "heart" in Chinook. Sure enough, our hearts are here, and always will be.

But down the road, we found a little piece of land where we can build our dream house, the one we've planned for years. I'll use my cordless drill and cordless saw to build it, no doubt. And the Yankees will languish in the toolbox, awaiting someone's hand. It won't be mine. Ren doesn't have any interest in old tools, not after she used my cordless tools. They fascinate her, as much as they intrigue me. How long will they last? Long enough to hand them down? Or will they be obsolete overnight, when some new advance comes along?

I don't know who will own, use, and love my Yankees when I'm gone, or any of the other old hand tools in this book. Maybe you. They'll make fine conversation pieces and valuable antiques, if that's their destiny. But they all function as well after several working lifetimes and many owners as when they were new. Watch the garage sales.

The major advances in civilization are processes that all but wreck the societies in which they occur.

ALFRED NORTH WHITEHEAD

JEFF TAYLOR has been a professional carpenter for twenty years. He contributes a monthly column, "Toolbox," to *Harrowsmith Country Life*. He lives in a small town in Oregon where he is building a new house by hand.

RICH IWASAKI is a location photographer based in Portland, Oregon. The images of tools that appear in this book are his first venture into photography of subjects that don't move.

As collaborators, Jeff and Rich have published articles in *Mother Earth News*, *Harrowsmith Country Life*, *Writer's Digest*, and other magazines. They have been working together for ten years. This is the first book for each of them.